Lecture Notes in Control and Information Sciences

Edited by M. Thoma

Vol. 43: Stochastic Differential Systems
Proceedings of the 2nd Bad Honnef Conference
of the SFB 72 of the DFG at the University of Bonn
June 28 – July 2, 1982
Edited by M. Kohlmann and N. Christopeit
XII, 377 pages. 1982.

Vol. 44: Analysis and Optimization of Systems
Proceedings of the Fifth International
Conference on Analysis and Optimization of Systems
Versailles. December 14–17, 1982
Edited by A. Bensoussan and J. L. Lions
XV, 987 pages, 1982

Vol. 45: M. Arató
Linear Stochastic Systems
with Constant Coefficients
A Statistical Approach
IX, 309 pages. 1982

Vol. 46: Time-Scale Modeling of Dynamic Networks
with Applications to Power Systems
Edited by J. H. Chow
X, 218 pages. 1982

Vol. 47: P. A. Ioannou, P. V. Kokotovic
Adaptive Systems with Reduced Models
V, 162 pages. 1983

Vol. 48: Yaakov Yavin
Feedback Strategies for Partially
Observable Stochastic Systems
VI, 233 pages, 1983

Vol. 49: Theory and Application of Random Fields
Proceedings of the IFIP-WG 7/1
Working Conference
held under the joint auspices of the
Indian Statistical Institute
Bangalore, India, January 1982
Edited by G. Kallianpur
VI. 290 pages. 1983

Vol. 50: M. Papageorgiou
Applications of Automatic Control Concepts
to Traffic Flow Modeling and Control
IX, 186 pages. 1983

Vol. 51: Z. Nahorski, H.F. Ravn, R.V.V. Vidal
Optimization of Discrete Time Systems
The Upper Boundary Approach
V, 137 pages 1983

Vol. 52: A. L. Dontchev
Perturbations, Approximations and Sensitivity Analysis
of Optimal Control Systems
IV, 158 pages. 1983

Vol. 53: Liu Chen Hui
General Decoupling Theory of Multivariable
Process Control Systems
XI, 474 pages. 1983

Vol. 54: Control Theory for Distributed
Parameter Systems and Applications
Edited by F. Kappel, K. Kunisch,
W. Schappacher
VII, 245 pages. 1983.

Vol. 55: Ganti Prasada Rao
Piecewise Constant Orthogonal Functions
and Their Application to Systems and Control
VII, 254 pages. 1983.

Vol. 56: Dines Chandra Saha, Ganti Prasada Rao
Identification of Continuous
Dynamical Systems
The Poisson Moment Functional
(PMF) Approach
IX, 158 pages. 1983.

Vol. 57: T. Söderström, P. G. Stoica
Instrumental Variable Methods
for System Identification
VII, 243 pages. 1983.

Vol. 58: Mathematical Theory of
Networks and Systems
Proceedings of the MTNS-83 International
Symposium
Beer Sheva, Israel, June 20–24, 1983
Edited by P. A. Fuhrmann
X, 906 pages. 1984

Vol. 59: System Modelling and Optimization
Proceedings of the 11th IFIP Conference
Copenhagen, Denmark, July 25-29, 1983
Edited by P. Thoft-Christensen
IX, 892 pages. 1984

Vol. 60: Modelling and Performance
Evaluation Methodology
Proceedings of the International Seminar
Paris, France, January 24–26, 1983
Edited by F. Bacelli and G. Fayolle
VII, 655 pages. 1984

Vol. 61: Filtering and Control of Random
Processes
Proceedings of the E.N.S.T.-C.N.E.T. Colloquium
Paris, France, February 23–24, 1983
Edited by H. Korezlioglu, G. Mazziotto, and
J. Szpirglas
V, 325 pages. 1984

For information about Vols. 1–42 please contact your bookseller or Springer-Verlag.

Lecture Notes in Control and Information Sciences

Edited by M. Thoma and A. Wyner

99

S. P. Bhattacharyya

Robust Stabilization Against Structured Perturbations

Springer-Verlag
Berlin Heidelberg GmbH

Author
Shankar P. Bhattacharyya
Department of Electrical Engineering
Texas A & M University
College Station
Texas 77843
USA

ISBN 978-3-540-18056-2

Library of Congress Cataloging in Publication Data
Battacharyya, S. P. (Shankar P.),
Robust stabilization against structured perturbations.
(Lecture notes in control and information sciences; 99)
Bibliography: p.
1. Perturbation (Mathematics)
2. Control theory. 3. System design.
I. Title.
II. Series.
QA871.B47 1986 515.3'53 87-16515
ISBN 978-3-540-18056-2 ISBN 978-3-540-47729-7 (eBook)
DOI 10.1007/978-3-540-47729-7

Originally published by Springer-Verlag Berlin Heidelberg New York in 1987

2161/3020-543210

This monograph is dedicated

to my parents

and to my brother

Tarak P. Bhattacharyya

PREFACE

This book deals with the analysis and design of control systems for plants which contain physical parameters subject to perturbation. The physical parameters could consist of masses, inertias, spring constants, aerodynamic coefficients etc., that are required in the mathematical description of the dynamics of the plant. In engineering, one frequently encounters situations where the structure of the plant and the nominal values of these parameters are known, but the parameters undergo large perturbations as the operating conditions of the control system change. This problem cannot be treated within the framework of the familiar theory of robust control where transfer function norms are used to describe the plant perturbation class. The latter class of perturbations is unstructured as opposed to the highly structured class of perturbations that is relevant here.

We consider linear time invariant systems and focus on the problem of closed loop stability under perturbations of a real parameter vector representing the physical parameters. Our objectives are a) to analyze the stability of the closed loop system for prescribed ranges of perturbation b) to estimate the size of the stability region in this parameter space as a function of controller parameters and c) to thereby design controllers that provide adequate stability margins.

Solutions to the above problems are developed in the transfer function and state space domains at both the theoretical and the algorithmic, computational levels. Some auxiliary, related problems, dealing with feedback stabilization with controllers of low dynamic order are also considered.

The results described in the book were mostly obtained by the author and his coworkers in the last two years. I would like to thank Leehyun Keel for several ideas that appear in Chapters 4-7, and for doing most of the computational work. I am grateful to Radek Biernacki for collaborating with me on work leading to the results of Chapters 2 and 3. Humor S. Hwang did the examples in Chapter 3. I thank Bob Barmish and R.K.Yedavalli for several useful discussions on structured perturbations, and John Fleming for suggesting many improvements to the initial draft of the manuscript. It is a pleasure to acknowledge the support and encouragement of my longstanding friend and colleague, Jo Howze. As for Boyd Pearson, it is impossible to thank him for everything he has taught me. I acknowledge the National Science Foundation's financial support of this research. I am grateful to Didi who urged me to write this book. She and Supriya provided solid support at crucial periods over the last two years for which I am thankful. Finally I am very grateful to Mary D. Sehlhoff for her expert typing of the manuscript.

March 23,1987 S.P. Bhattacharyya

College Station, Texas

TABLE OF CONTENTS

CHAPTER 1

BACKGROUND AND PRELIMINARIES

1. INTRODUCTION

The design of a control system is invariably based on an assumed nominal model of the plant to be controlled. After the usual simplifications, such as linearization about an operating point, lumped parameter approximation etc., one ends up most often with a linear time invariant system described by a prescribed set of differential equations for the nominal model of the plant. At this point, various established design strategies such as the frequency domain methods of classical control, or the state space methods of optimal control theory can be applied to the nominal model to produce a feedback controller that yields closed loop stability and an acceptable output time response.

An important and fundamental practical requirement to be satisfied by the controller is the invariance of the property of closed loop stability, under perturbations, from a suitable class, of the nominal plant model. A controller satisfying this requirement is said to be robust with respect to the prescribed class of perturbations. The theory of analysis and design of such controllers is currently an active area of research in control theory (see for example, [1]-[43] and references cited therein.)

The specific description of the class of perturbations against which robustness is required depends, of course, on the physics and engineering of the particular plant in question. The general theory, however, distinguishes broadly between two types of perturbation classes-structured and unstructured.

2. STRUCTURED AND UNSTRUCTURED PERTURBATIONS

In the traditional unstructured approach [1]-[4] to dealing with perturbations, the nominal plant is represented by the transfer function matrix $\mathbf{G}_o(s)$ and the perturbed plant by the transfer function matrix $\mathbf{G}(s) = \mathbf{G}_o(s) + \Delta\mathbf{G}(s)$. The class of perturbations to be handled is described by assuming that, for a given stable proper rational function $\mathrm{R}(s)$

$$||\Delta\mathbf{G}(j\omega)|| \leq |\mathrm{R}(j\omega)|, \qquad \forall \omega \epsilon R \tag{2.1}$$

where

$$||\Delta\mathbf{G}(j\omega)|| \ := \bar{\sigma}[\Delta\mathbf{G}(j\omega)] \tag{2.2}$$

and $\bar{\sigma}[\cdot]$ denotes the maximum singular value.

Suppose now that $\mathbf{C}(s)$ is the transfer function of a feedback controller that stabilizes the nominal plant $\mathbf{G}_o(s)$. The main result of the theory of unstructured robust stability is a necessary and sufficient condition for $\mathbf{C}(s)$ to stabilize the entire family of perturbed plants described by (2.1). This result states that, under some mild technical assumptions, $\mathbf{C}(s)$ stabilizes all perturbed plants determined by (2.1) iff

$$||\mathbf{C}(\mathbf{I} + \mathbf{G}_o\mathbf{C})^{-1}(j\omega)|| \cdot |\mathrm{R}(j\omega)| \leq 1 \ \ \forall \omega \epsilon R \ . \tag{2.3}$$

The above result, which was proved in [1], is useful for checking the robustness of a given controller. The condition (2.3) has also been used in [2] to develop a synthesis procedure in the case of single input single output systems. A similar condition has been stated in [1] for norm bounded multiplicative perturbations.

The class of perturbations described by (2.1) is unstructured in the sense that the norm bound (2.1) allows perturbations $\Delta\mathbf{G}(s)$ to occur in "all directions" in the appropriate space of transfer functions. It is our view that many engineering problems cannot be dealt with adequately using this approach. This is due to the fact that the dynamical equations of most engineering plants, such as aircraft, robots, and chemical processes are usually known. Thus, good mathematical models are available and the system structure is well known qualitatively but there exists uncertainty regarding the numerical values of various physical parameters in the model. Spring constants, masses and inertias, reaction rates, and aerodynamic coefficients are common examples of such parameters. The uncertainty in turn may be due to the inability to measure various physical quantities, actual variations of parameters due to aging or to changes in the operating conditions of the system. There also exist uncertainties or errors in the modelling process which take the form of changes in transfer function coefficients or perturbation of the state space matrices. Such perturbations are only remotely related to any transfer function norm. In fact the class of unstructured perturbations determined by a transfer function norm bound generates a very rich class of systems. Design based on protection against such a large class of perturbations may result in very conservative systems when only physical parameters are subject to perturbation. For these reasons, there is a growing interest (see [5]-[43]) in the structured perturbation robust control problem. The distinct approaches that are developing may be classified into the polynomial approach [5]-[23], the Lyapunov or state space based approaches, [22]-[36], the μ synthesis approach [37]-[40], and the multi model simultaneous stabilization approach [41],[42].

In this monograph, we present some new results on this problem using the transfer function (Hurwitz) and the state space (Lyapunov) approaches. It will be assumed that the plant transfer function matrix $\mathbf{G}(s)$ or the plant state space model is dependent on a real parameter vector $\mathbf{p}$ representing physical parameters with nominal value $\mathbf{p}^0$ and $\Delta\mathbf{p} := \mathbf{p} - \mathbf{p}^0$ represents a perturbation. Clearly, larger (smaller) values of $||\Delta\mathbf{p}||_2$ correspond to larger (smaller) perturbations. For a given stabilizing controller there exists a largest value $\rho(\mathbf{p}^0)$ of $||\Delta\mathbf{p}||_2$ for which closed loop stability is preserved. This value therefore serves as a measure of stability margin. Based on these considerations we formulate several problems to be solved in the next few chapters.

Problem A Determining the Largest Stability Hypersphere

For a given stabilizing controller $\mathbf{C}(s)$ determine the radius $\rho(\mathbf{p}^0)$ of the stability hypersphere centered at $\mathbf{p}^0$ defined by the condition that whenever $||\Delta\mathbf{p}||_2 < \rho(\mathbf{p}^0)$ the closed loop system with plant parameter $\mathbf{p}^0 + \Delta\mathbf{p}$ is stable and there exists at least one perturbation $\overline{\Delta\mathbf{p}}$ with $||\overline{\Delta\mathbf{p}}||_2 = \rho(\mathbf{p}^0)$ such that the closed loop system with the parameter $\mathbf{p}^0 + \overline{\Delta\mathbf{p}}$ is not stable.

Problem B Robust Controller Design

Let $\mathbf{C}(s)$ denote a stabilizing controller with an adjustable parameter vector $\mathbf{x} \in R^s$. Determine a procedure to choose $\mathbf{x}$ so that the radius $\rho(\mathbf{p}^0, \mathbf{x})$ of the stability hypersphere is increased as a function of $\mathbf{x}$ until it contains a given class of perturbations $\{\Delta\mathbf{p}\}$.

Solutions to the above two problems will be developed in Chapters 2-5. These solutions will also allow us to treat the following special types of perturbation classes :

(i) Δp_i, the ith component of $\Delta\mathbf{p}$, is bounded by

$$-\gamma_i < \Delta p_i < \epsilon_i \tag{2.7}$$

for given positive numbers γ_i, ϵ_i, $i = 1, \ldots, k$.

(ii) The perturbation bounds are given by

$$-w_i\epsilon < \Delta p_i < w_i\epsilon \ , \quad i = 1, \ldots, k \tag{2.8}$$

where w_i are weights and ϵ is a positive constant.

A natural solution to the problem with the above perturbation classes will be obtained within the framework of our approach by determining the largest stability <u>ellipsoid</u> in parameter space.

The justification for using $\rho(\mathbf{p}^0, \mathbf{x})$ as a stability margin is that if $\mathbf{x}_1$ and $\mathbf{x}_2$ are two controllers with

$$\rho(\mathbf{p}^0, \mathbf{x}_1) > \rho(\mathbf{p}^0, \mathbf{x}_2) \tag{2.9}$$

then clearly $\mathbf{x}_1$ is "more robust" than $\mathbf{x}_2$ because the corresponding stability hypersphere denoted by $\mathbf{S}_\rho(\mathbf{p}^0, \mathbf{x}_1)$ is larger than $\mathbf{S}_\rho(\mathbf{p}^0, \mathbf{x}_2)$ and in fact

$$\mathbf{S}_\rho(\mathbf{p}^0, \mathbf{x}_2) \subset \mathbf{S}_\rho(\mathbf{p}^0, \mathbf{x}_1) \tag{2.10}$$

so that the family of perturbed plants guaranteed to be stabilized by $\mathbf{x}_1$ contains the family that is guaranteed to be stabilized by $\mathbf{x}_2$.

Generally speaking, the problem described above deals with the preservation of the Hurwitz property (i.e. roots in the open left half plane) of a set of polynomials generated by perturbing the coefficients of a nominal polynomial. We proceed to describe, in the next section, two important recent results on this problem.

3. HURWITZ REGIONS IN COEFFICIENT SPACE

Let

$$\delta^o(s) = \delta_o^o + \delta_1^0 s + \cdots + \delta_n^o s^n \tag{3.1}$$

denote a Hurwitz polynomial that represents for example, the characteristic polynomial of a closed loop control system for nominal values of the plant and controller parameters. Under plant parameter perturbations the coefficients of this polynomial change and it is of interest to determine a) whether all the polynomials obtained by perturbing each coefficient within a prescribed interval are Hurwitz and b) the size of the largest stability hypersphere in this coefficient space centered at the nominal coefficient vector. The first of these problems was solved in [5] and has come to be known as Kharitonov's theorem. The second problem was solved in [6]. We state both results below.

Consider the family of polynomials $\{\delta(s)\}$ given by

$$\delta(s) = \delta_o + \delta_1 s + \cdots + \delta_n s^n \tag{3.2a}$$

with each coefficient varying continuously within a prescribed interval:

$$x_i \leq \delta_i \leq y_i, \; i = 0,\ 1, \cdots n. \tag{3.2b}$$

<u>**Theorem 3.1 (Kharitonov's Theorem)**</u>

<u>The family of polynomials (3.2) are all Hurwitz if and only if the four polynomials</u>

$$\underline{\delta^1(s), \delta^2(s), \delta^3(s) \text{ and } \delta^4(s)}$$

<u>given below are Hurwitz:</u>

$$\delta^1(s) = y_o + x_1 s + x_2 s^2 + y_3 s^3 + y_4 s^4 + x_5 s^5 + x_6 s^6 + \cdots \tag{3.3a}$$

$$\delta^2(s) = y_o + y_1 s + x_2 s^2 + x_3 s^3 + y_4 s^4 + y_5 s^5 + x_6 s^6 + \cdots \tag{3.3b}$$

$$\delta^3(s) = x_o + x_1 s + y_2 s^2 + y_3 s^3 + x_4 s^4 + x_5 s^5 + y_6 s^6 \cdots \tag{3.3c}$$

$$\delta^4(s) = x_o + y_1 s + y_2 s^2 + x_3 s^3 + x_4 s^4 + y_5 s^5 + y_6 s^6 + \cdots \tag{3.3d}$$

This powerful theorem was first proved in [5]. An alternative system theoretic proof was given in [7]. It is emphasized that the theorem reduces the formidable task of verifying the Hurwitz property of the infinite family of polynomials (3.2) to the simple one of checking the stability of the four prescribed "extreme" polynomials (3.3). An independent, simple proof of this theorem, due to Hervé Chapellat [69], is given in Section 6. This proof is similar to Kharitonov's but illustrates more clearly the intuitive content of the theorem.

The above theorem is useful for analysis because it solves the problem of determining the stability of the vectors $\delta^o + \Delta\delta$ as $\Delta\delta$ varies within a prescribed hypercube. For synthesis and design problems it is of importance to find out the largest such stability hypercube but this is a difficult unsolved problem.

On the other hand, the problem of determining the largest stability hypersphere in the space δ has been recently solved in [6]. To state this important result let

$$\Delta_o := \{\delta | \delta \epsilon R^{n+1}, \delta_o = 0\} \tag{3.4a}$$

$$\Delta_n := \{\delta | \delta \epsilon R^{n+1}, \delta_n = 0\} \tag{3.4b}$$

and for any real ω

$$\Delta(\omega) := \{\delta | \delta \epsilon R^{n+1}, \delta(s) = (s^2 + \omega^2)\ell(s), \quad \ell(s) \text{ arbitrary}\}. \tag{3.4c}$$

Note that Δ_o, Δ_n and $\Delta(\omega)$ are subspaces of R^{n+1} and correspond respectively, to polynomials $\delta(s)$ with roots at $s = 0$, $s = \infty$, and $s = \pm j\omega$. Let d_o, d_n and $d(\omega)$ denote the Euclidean distances between the nominal vector δ^o and the subspaces Δ_o, Δ_n and $\Delta(\omega)$

respectively.

Let

$$d := \inf_{0 \leq \omega \leq \infty} d(\omega). \tag{3.5}$$

Theorem 3.2

The radius of the largest stability hypersphere in the coefficient space δ centered at δ^o is given by

$$\rho(\delta^o) = \min\{d_o, d_n, d\}. \tag{3.6}$$

This theorem was proved in [6] where formulas for the calculation of d_0, d_n and $d(\omega)$ were also given. The result is nice because it gives a simple calculation to estimate the size of the Hurwitz region in the coefficient space δ by fitting the largest stable hypersphere centered at δ^o into this region. In Chapter 2, we solve a more useful and general version of this problem applicable to the plant parameter space.

4. STATE SPACE PERTURBATIONS

The state space description of dynamic systems are often based on the choice of physical quantities as state variables. In this description the matrices representing the model contain various physical parameters as entries. The robust stability and stabilization problems can be formulated more meaningfully in the state space, in such cases.

Let

$$\dot{\mathbf{x}}(t) = \mathrm{A}^\circ\ \mathbf{x}(t) \tag{4.1}$$

represent the dynamics of a stable system. Under parameter perturbations A° changes to $\mathrm{A}^\circ + \Delta\mathrm{A}$. The problem of determining bounds on the elements of $\Delta\mathrm{A}$ that guarantee stability of the matrices $\mathrm{A}^\circ + \Delta\mathrm{A}$ has been studied in the literature using Gershgorin's theorem [36] and also Lyapunov methods [24]-[28]. Patel and Toda [28] have given the following result:

Theorem 4.1 [28]

If A^0 is stable $A^0 + \Delta A$ remains stable if the elements ΔA_{ij} of the $n \times n$ matrix ΔA satisfy

$$|\Delta A_{ij}| < \frac{1}{n\sigma_{\max}(P)} \tag{4.2}$$

where P is the unique positive definite solution of the Lyapunov equation

$$A^T P + PA + 2I_n = 0 \tag{4.3}$$

A structured version of this bound has been given by Yedavalli in [24], where the right hand side of (4.2) was replaced by

$$\frac{1}{\sigma_{\max}(|P|U_n)_s} \tag{4.4}$$

where U_n is an $n \times n$ matrix whose entries are 0 or 1 corresponding to fixed or variable entries respectively, in A, $|P|$ denotes the matrix P with each entry replaced by its absolute value, and $(\cdot)s$ denotes the symmetric part of the matrix in brackets.

In the control problem the matrix A will represent the closed loop system containing plant and controller. Moreover, instead of arbitrary perturbations, the structure of the plant must be considered. In fact, the results of [43] have shown that any conclusion drawn by allowing arbitrary perturbations of state space models can be very misleading. Thus, certain elements of A must remain fixed and perturbation free either by definition of the variables or because they originate from controller parameters that are fixed. To synthesize the controller the perturbation bounds should be determined as a function of controller parameters so that they may be enlarged by appropriate choice of these parameters. This problem is treated in Chapter 5, using the Lyapunov based approach.

5. DISCUSSION OF CONTENTS

The results described in sections 3 and 4 above are elegant, mathematically, but are not adequate to deal with the control problem, as they ignore all structural information about the plant and controller. This, in turn, is due to the assumption in both Theorems 3.1 and 3.2 that the characteristic polynomial coefficients perturb independently. Likewise, in Theorem 4.1 all elements of A are subject to perturbation. For the control problem, the closed loop characteristic polynomial coefficients are functions of the controller and plant parameter vectors and only the latter is subject to perturbations. Kharitonov's theorem, unfortunately, does not generalize in an obvious manner to the space of plant parameters. Despite this drawback, the same type of problem has been treated in several recent papers, on the stability of sets of polynomials, (see [5]-[16]). Of these papers [8]-[14] deal with independent perturbations of the coefficients of the closed loop characteristic polynomial and various sufficient conditions for stability are given. In [13] and [14] the geometry of Hurwitz polynomials for discrete systems systems has been studied, again in the space of coefficients of the closed loop characteristic polynomial.

The maximization of the general "box-type" kind of perturbations described in (2.7) has been dealt with in [16]. A special case of (2.8) was treated in [15] where **p** was considered to be the closed loop characteristic polynomial coefficient vector and Kharitonov's theorem was used to determine the maximum value of ϵ. However, these results are necessarily conservative because the coefficients are, once again, assumed to perturb independently.

These considerations motivate us to formulate the problem of calculating the stability margins and stability regions directly in terms of the parameter **p**. In Chapter 2, we calculate the largest stability hypersphere in the parameter space **p** under the assumption that **p** enters linearly or affinely into the plant characteristic polynomial coefficients. In Chapter 3, under the same assumption, we determine conditions for stability under the polytope of perturbations (2.7). From the point of view of analysis (i.e. controller is given) these results will constitute a generalization of Theorems 3.1 and 3.2 of this chapter.

The synthesis problem will be treated by us throughout, under the assumption that a real vector **x** of fixed dimension completely parametrizes the controller.This corresponds to fixing the controller order. The design problem in robust stabilization is to choose **x** so that stability is guaranteed for prescribed ranges of parameter perturbations. The solution of the design problem given in Chapters 2 and 3 is facilitated by the fact that the stability regions are calculated explicitly in terms of controller parameters. The results of Chapters 2 and 3 are based on the report [20].

In Chapter 4, the general case of nonlinear dependence of the characteristic polynomial coefficients on the parameter is considered. The results of Chapter 2 are applied to this problem after a preliminary calculation of the characteristic polynomial in a linear separated form. In this case, the stability margin calculated may be conservative since it does not correspond to the largest stability domains. These results extend the earlier results of [18] and [19].

Chapter 5 presents a state space treatment of the structured robust stability and stabilization problems. The Lyapunov based approach of [24] and [28] is extended to take the structure of the plant perturbations into account and to define a stability margin as a function of the controller parameters. The design or robustification algorithm gives a numerical procedure to increase this margin. The results of Chapter 4 and 5 were originally reported in [23].

The procedures developed in Chapters 2-5 are applicable once a stabilizing controller has been found. An important practical consideration in controller design is that the design parameter vector that yields robust stability should not be of unduly high dimension as this bogs down the subsequent design process where adjustments to the controller must be made to attain other objectives. This motivates us to develop some results on stabilization with fixed or low order controllers. In Chapter 6, we use the characteristic polynomial formulation and results from linear programming and the stability hypersphere calculation of Theorem 3.2 to give new bounds on the order of stabilizing controllers. In Chapter 7, an algorithm is developed for stabilization with a controller of low dynamic order, using the Sylvester equation formulation of [44][45] and [46] extended to the case of dynamic compensation. These results were first reported in [47]. The contents of Chapters 4-7 were developed as part of the Ph.D. thesis of L.H. Keel [22]. Chapter 8 summarises the results and discusses future research directions.

In Chapters 3-7, we have included some numerical examples to illustrate the design procedures. The examples in Chapter 3 were solved by Humor S. Hwang [17]. The algorithms and example problems in Chapters 4-7 were programmed by L.H.Keel [22]. Many of these involve the optimization of a function (for example, stability margin) with respect to several variables (say, controller parameters). These have all been executed using the standard optimization routines available in the Harwell library [48]. Whenever possible the gradients of the objective functions have been explicitly calculated in the text. In each of the cases considered, however, nothing is known about the geometry of the objective function and all that can be said is that the algorithms provide descent or ascent procedures to find local optima. This is not usually a serious limitation because these algorithms are to be viewed merely as devices to automate computational procedures to sequentially improve a given design until a satisfactory controller is obtained.

We emphasize that the results described here are necessarily preliminary because the problems treated here do not deal with performance measures related to system response such as tracking [49], disturbance rejection and transient response [50],[51]. Their aim, however is to provide effective and simple solutions to the problem of obtaining robust stability against real plant parameter variations occuring in prescribed ranges, with controllers of low dynamic order. Although this is an essential first step in many design procedures, it has received very little attention in the control theory literature. We expect that the results given here will provide some helpful tools to the practicing control engineer and will also spur the development of a more complete and effective theory of design.

Convention for references: In the rest of the book we have adopted the following convention to refer to equations, theorems and lemmas: A reference to equation (4.3) in a given chapter denotes equation (4.3) in Section 4 of the same chapter, while Theorem 6.5.4 denotes Theorem 5.4 in Chapter 6, Section 5.

6. PROOF OF KHARITONOV'S THEOREM

The proof depends on the well known interlacing property of the roots of the odd and even parts of stable (i.e. strictly Hurwitz) polynomials, given, for instance, in [70, pp. 271]. Let $p(s)$ be an arbitrary polynomial and write

$$p(s) = \underbrace{p_e(s)}_{\text{even degree terms}} + \underbrace{p_o(s)}_{\text{odd degree terms}} \tag{6.1}$$

Theorem 6.1[70]

The polynomial $p(s)$ is stable if and only if the leading coefficients of $p_e(s)$ and $p_o(s)$ are of the same sign and the roots of $p_e(s)$ and $p_o(s)$ all lie on the imaginary axis, are nonrepeated, and interlace:

$$\cdots < -\omega_{o1} < -\omega_{e1} < 0 < \omega_{e1} < \omega_{o1} < \omega_{e2}, \cdots$$

where $\pm j\omega_{ei}$ are the roots of $p_e(s) = 0$ and 0, $\pm j\omega_{oi}$ are the roots of $p_o(s) = 0$.

This theorem is illustrated in Figure 6.1. We shall say loosely, that $p_e(s)$ and $p_o(s)$ interlace if they satisfy the conditions of this theorem. We will also need the following lemmas.

Lemma 6.2

Let

$$p_1(s) = p_e(s) + p_{o1}(s) \tag{6.2a}$$

$$p_2(s) = p_e(s) + p_{o2}(s) \tag{6.2b}$$

denote two stable polynomials of the same degree with the same even part $p_e(s)$ and differing odd parts $p_{o1}(s)$ and $p_{o2}(s)$ satisfying

$$\frac{p_{o1}(j\omega)}{j\omega} \leq \frac{p_{o2}(j\omega)}{j\omega} \quad \forall\, \omega \in [0, \infty] \tag{6.3}$$

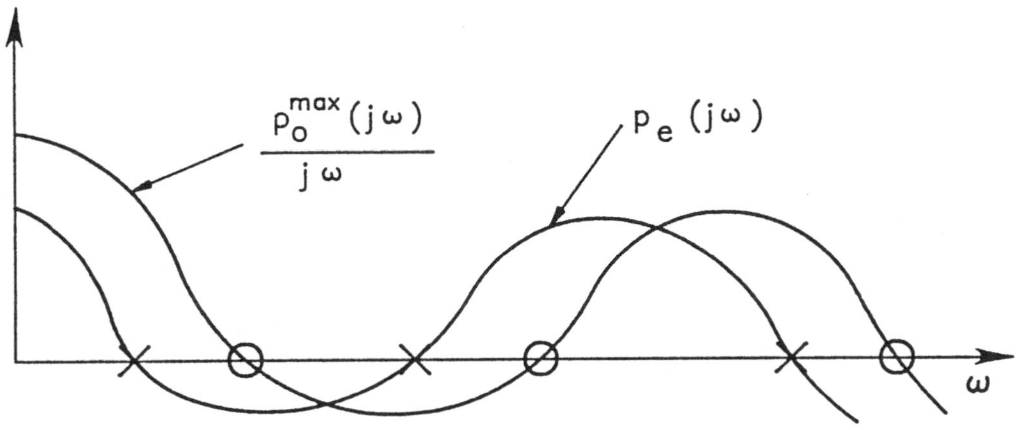

Illustration of Theorem 6.1

Figure 6.1

Then

$$p(s) = p_e(s) + p_o(s)$$

is stable for every $p_o(s)$ satisfying

$$\frac{p_{o1}(j\omega)}{j\omega} \leq \frac{p_o(j\omega)}{j\omega} \leq \frac{p_{o2}(j\omega)}{j\omega} \qquad \forall\ \omega \in [0, \infty]. \tag{6.4}$$

Proof

Since $p_1(s)$ and $p_2(s)$ are stable, $p_{o1}(s)$ and also $p_{o2}(s)$ interlaces with $p_e(s)$. It is easy to show that $p_o(s)$ has the same degree as $p_{o1}(s)$ and $p_{o2}(s)$, and that their leading coefficients all have the same sign. Then the condition (6.4) forces $p_o(s)$ to interlace with $p_e(s)$. Therefore, $p_e(s) + p_o(s) = p(s)$ is stable. ◇

This lemma is illustrated in Figure 6.2.

The dual of this result is stated next, without proof.

Lemma 6.3

Let

$$p_1(s) = p_{e1}(s) + p_o(s) \tag{6.5a}$$

$$p_2(s) = p_{e2}(s) + p_o(s) \tag{6.5b}$$

denote two stable polynomials of the same degree with the same odd part $p_o(s)$ and differing even parts $p_{e1}(s)$ and $p_{e2}(s)$ satisfying

$$p_{e1}(j\omega) \leq p_{e2}(j\omega) \qquad \forall\ \omega \in [0,\ \infty]. \tag{6.6}$$

Then

$$p(s) = p_e(s) + p_o(s)$$

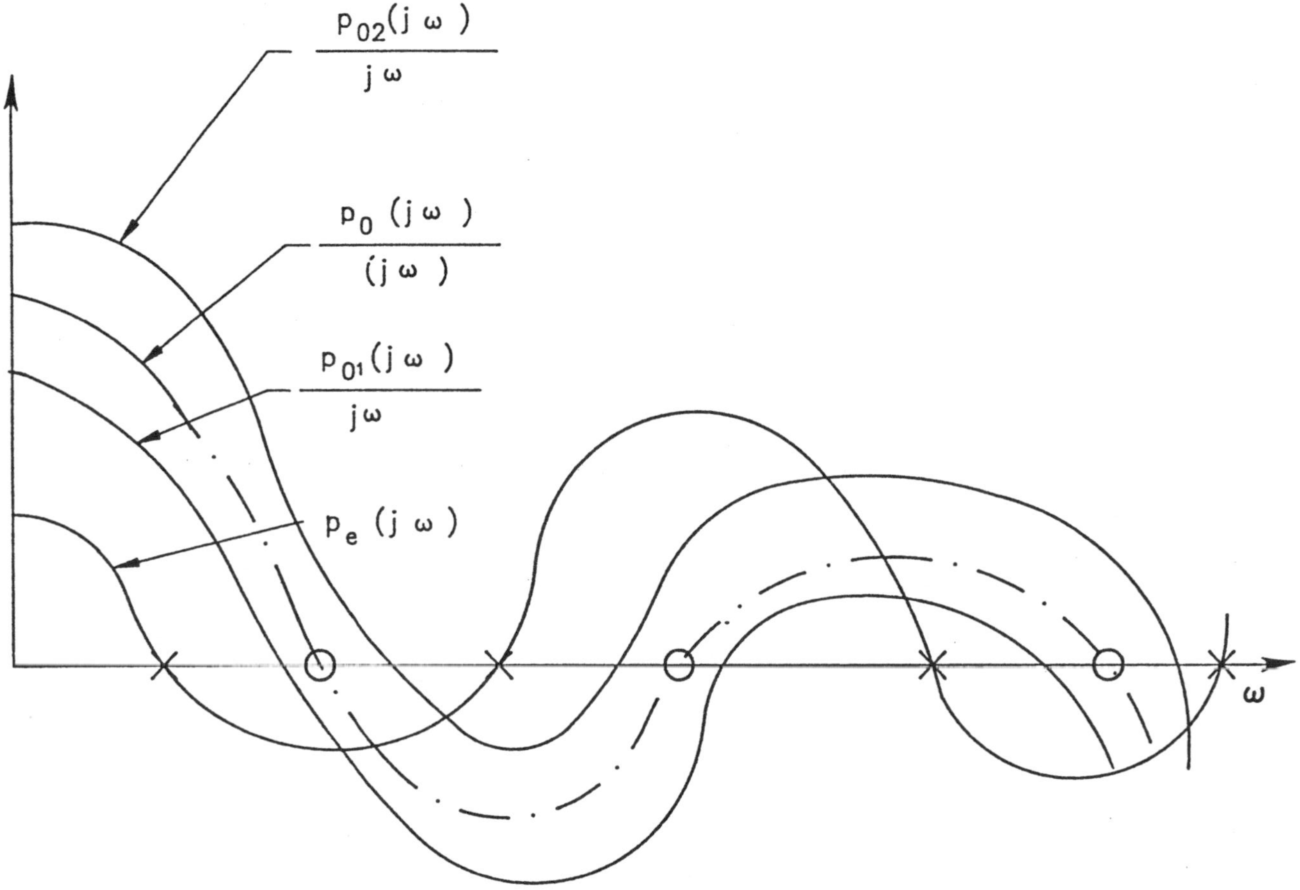

Illustration of Lemma 6.2

Figure 6.2

is stable for every $p_e(s)$ satisfying

$$p_{e1}(j\omega) \leq p_e(j\omega) \leq p_{e2}(j\omega) \quad \forall \quad \omega \in [0, \infty]. \tag{6.7}$$

We are now ready to prove Kharitonov's theorem. Let us introduce the box $\mathcal{B}$ of coefficients of the perturbed polynomials:

$$\mathcal{B} := \{\delta | \delta \in R^{n+1}, x_i \leq \delta_i \leq y_i, i = 0, 1, \cdots, n\}. \tag{6.8}$$

The Kharitonov polynomials (3.3) repeated below, for convenience are:

$$\delta^1(s) = y_o + x_1 s + x_2 s^2 + y_3 s^3 + y_4 s^4 + x_5 s^5 + x_6 s^6 + \cdots \tag{6.9a}$$

$$\delta^2(s) = y_o + y_1 s + x_2 s^2 + x_3 s^3 + y_4 s^4 + y_5 s^5 + x_6 s^6 + \cdots \tag{6.9b}$$

$$\delta^3(s) = x_o + x_1 s + y_2 s^2 + y_3 s^3 + x_4 s^4 + x_5 s^5 + y_6 s^6 + \cdots \tag{6.9c}$$

$$\delta^4(s) = x_o + y_1 s + y_2 s^2 + x_3 s^3 + x_4 s^4 + y_5 s^5 + y_6 s^6 + \cdots \quad . \tag{6.9d}$$

These polynomials are built from two different even parts $p_e^{\max}(s)$ and $p_e^{\min}(s)$ and two different odd parts $p_o^{\max}(s)$ and $p_o^{\min}(s)$ defined below:

$$p_e^{\max}(s) := y_o + x_2 s^2 + y_4 s^4 + x_6 s^6 + y_8 s^8 + \cdots \tag{6.10a}$$

$$p_e^{\min}(s) := x_o + y_2 s^2 + x_4 s^4 + y_6 s^6 + x_8 s^8 + \cdots \tag{6.10b}$$

and

$$p_o^{\max}(s) := y_1 s + x_3 s^3 + y_5 s^5 + x_7 s^7 + y_9 s^9 + \cdots \tag{6.10c}$$

$$p_o^{\min}(s) := x_1 s + y_3 s^3 + x_5 s^5 + y_7 s^7 + x_9 s^9 + \cdots \tag{6.10d}$$

The motivation for the superscripts max and min is as follows. Let $\delta(s)$ be an arbitrary polynomial with its coefficients lying in the box B and let $\delta^e(s)$ be its even part. Then

$$p_e^{\max}(j\omega) = y_0 - x_2\omega^2 + y_4\omega^4 - x_6\omega^6 + y_8\omega^8 \tag{6.11a}$$

$$\delta^e(j\omega) = \delta_0 - \delta_2\omega^2 + \delta_4\omega^4 - \delta_6\omega^6 + \delta_8\omega^8 \cdots \tag{6.11b}$$

$$p_e^{\min}(j\omega) = x_0 - y_2\omega^2 + x_4\omega^4 - y_6\omega^6 + x_8\omega^8 \cdots \tag{6.11c}$$

so that

$$p_e^{\max}(j\omega) - \delta^e(j\omega) = (y_0 - \delta_0) + (\delta_2 - x_2)\omega^2 + (y_4 - \delta_4)w^4 + (\delta_6 - x_6)\omega^6 + \cdots \tag{6.12a}$$

and

$$\delta^e(j\omega) - p_e^{\min}(j\omega) = (\delta_0 - x_0) + (y_2 - \delta_2)\omega^2 + (\delta_4 - x_4)\omega^4 + (y_6 - \delta_6)\omega^6 + \cdots \tag{6.12b}$$

Therefore,

$$p_e^{\min}(j\omega) \leq \delta^e(j\omega) \leq p_e^{\max}(j\omega) \quad \forall \quad \omega \in [0, \infty]. \tag{6.13}$$

Similarly, if $\delta^o(s)$ denotes the odd part of $\delta(s)$, it can be verified that

$$\frac{p_o^{\min}(j\omega)}{j\omega} \leq \frac{\delta^o(j\omega)}{j\omega} \leq \frac{p_o^{\max}(j\omega)}{j\omega} \quad \forall \ \omega \in [0, \infty]. \tag{6.14}$$

To proceed, note that the Kharitonov polynomials (6.9) can be rewritten as:

$$\delta^1(s) = p_e^{\max}(s) + p_o^{\min}(s) \tag{6.15a}$$

$$\delta^2(s) = p_e^{\max}(s) + p_o^{\max}(s) \tag{6.15b}$$

$$\delta^3(s) = p_e^{\min}(s) + p_o^{\min}(s) \tag{6.15c}$$

$$\delta^4(s) = p_e^{\min}(s) + p_o^{\max}(s) \tag{6.15d}$$

If all the polynomials with coefficients in the box B are stable, it is clear that the Kharitonov polynomials (6.9) must also be stable since their coefficients lie in B. For the converse, assume that the Kharitonov polynomials (6.9) are stable, and let $\delta(s) = \delta^e(s) + \delta^o(s)$ be an arbitrary polynomial with coefficients in the box B with even part $\delta^e(s)$ and odd part $\delta^o(s)$.

Since $\delta^1(s)$ and $\delta^2(s)$ are stable and (6.14) holds, we conclude, from Lemma 6.2 applied to $\delta^1(s)$ and $\delta^2(s)$ in (6.15a) and (6.15b), that

$$p_e^{\max}(s) + \delta^o(s) \text{ is stable.} \tag{6.16}$$

Similarly, from Lemma 6.2 applied to $\delta^3(s)$ and $\delta^4(s)$ in (6.15c) and (6.15d) we conclude that

$$p_e^{\min}(s) + \delta^o(s) \text{ is stable.} \tag{6.17}$$

Now, since (6.13) holds, applying Lemma 6.3 to the stable polynomials $p_e^{\max}(s) + \delta^o(s)$ and $p_e^{\min}(s) + \delta^o(s)$, we conclude that

$$\delta^e(s) + \delta^o(s) = \delta(s) \text{ is stable.}$$

This concludes the proof. $\diamond$.

The proof of Kharitonov's theorem is illustrated in Figure 6.3.

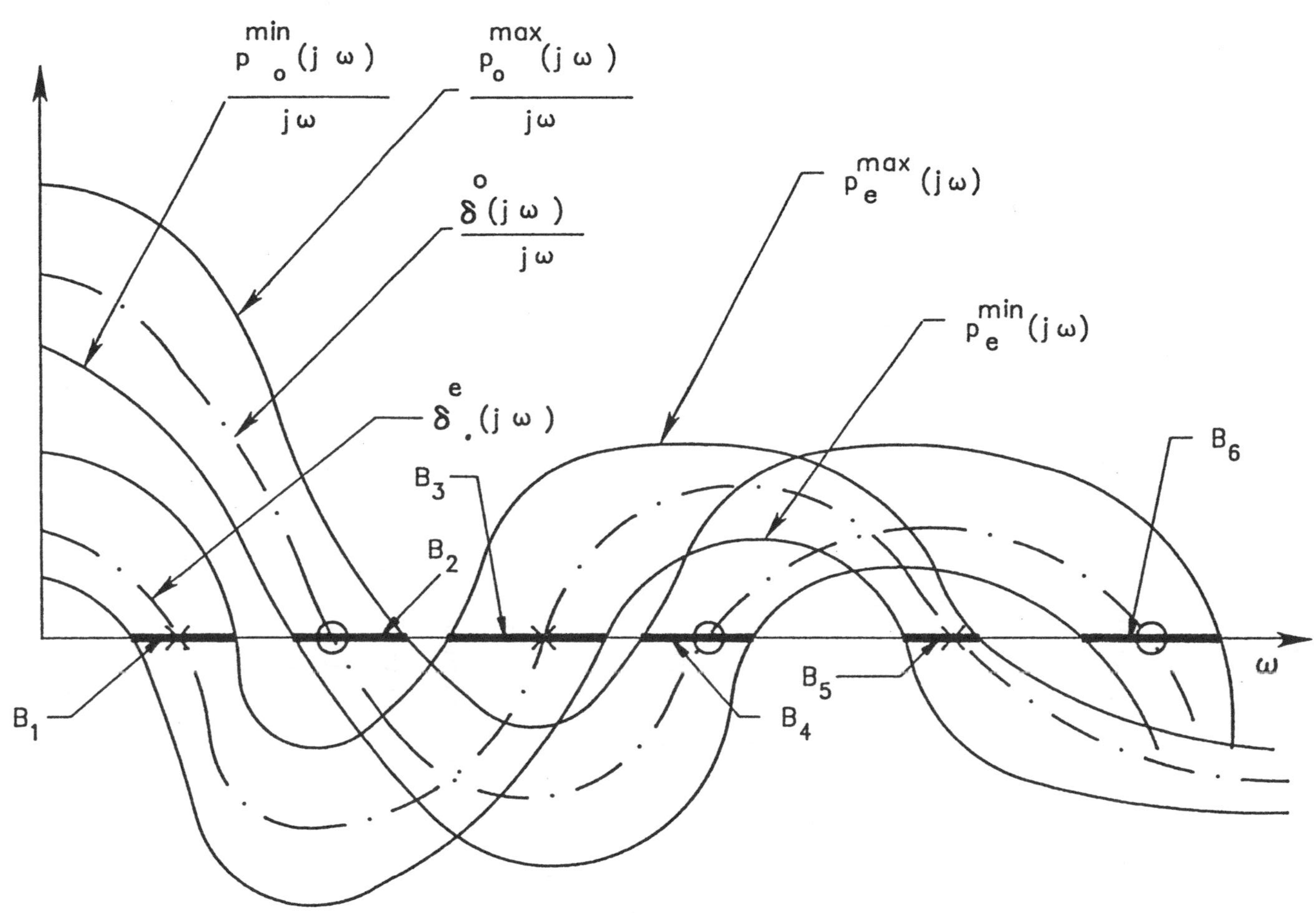

Illustration of Kharitonov's Theorem

Figure 6.3

The illustration shows how all polynomials with even parts bounded by $p_e^{\max}(s)$ and $p_e^{\min}(s)$ and odd parts bounded by $\frac{p_o^{\max}(s)}{s}$ and $\frac{p_o^{\min}(s)}{s}$ on the imaginary axis satisfy the interlacing property when the Kharitonov polynomials are stable. Figure 6.3 also shows that the interlacing property for the odd and even parts of a <u>single</u> stable polynomial, corresponding to a <u>point</u> δ in coefficient space generalizes to the <u>box</u> B of stable polynomials as the requirement of "interlacing" of the odd and even "<u>tubes</u>." This leads to the following alternative version of Kharitonov's theorem. Let $\pm\, j\omega_{e_i}^{\max}(\pm\, j\omega_{e_i}^{\min})$ denote the roots of $p_e^{\max}(s)(p_e^{\min}(s))$ and let $0, \pm j\omega_{o_i}^{\max}$ $(0, \pm\, j\omega_{o_i}^{\min})$ denote the roots of $p_o^{\max}(s)(p_o^{\min}(s))$.

<u>Theorem 6.4</u>

<u>The box B corresponds to stable polynomials if and only if</u>

$$0 < \omega_{e_1}^{\min} < \omega_{e_1}^{\max} < \omega_{o_1}^{\min} < \omega_{o_1}^{\max} < \omega_{e_2}^{\min} < \omega_{e_2}^{\max} < \omega_{o_2}^{\min} < \omega_{o_2}^{\max} < \cdots$$

This theorem states that stability of the polynomials in the box B corresponds to non overlapping of the frequency bands B_1, B_2, etc. shown in Figure 6.3. The implication of this fact on robust stability is that the perturbation box B can be enlarged until overlap occurs for some pair of adjacent bands.

CHAPTER 2

THE STABILITY HYPERSPHERE IN PARAMETER SPACE

1. INTRODUCTION

In the standard feedback system of Figure 2.1, suppose that the plant transfer function contains the parameter vector **p** and the controller is characterized by the real vector **x**, i.e.

$$\mathrm{G}(s) = \mathrm{G}(s, \mathbf{p})$$

$$\mathrm{C}(s) = \mathrm{C}(s, \mathbf{x})$$

and fixing the parameter vector **p** fixes the plant and the choice of the vector **x** is equivalent to picking the controller C(s).

The characteristic polynomial of the closed loop system is then given by

$$\delta(s, \mathbf{x}, \mathbf{p}) = \sum_{i=0}^{n} \delta_i(\mathbf{x}, \mathbf{p}) s^i. \tag{1.1}$$

In this chapter, we consider the case in which the $\delta_i(\mathbf{x},\ \mathbf{p})$ are linear or affine in **p** and for a fixed **x** calculate the radius of the largest stability hypersphere centered at $\mathbf{p}^0$. Although this is mathematically, a special case, it will <u>always</u> hold in single input (multioutput) or single output (multiinput) systems if the parameter vector **p** is taken to be the list of plant transfer function coefficients. This problem is formulated in Section 2 and solved in Section 3. We then extend our approach, in Section 4, to treat the case where the plant transfer coefficients are linearly or affinely dependent on some "primary parameters" and calculate the largest stability hypersphere in this space. This allows us to treat the practically important case where the transfer function coefficients are interdependent. It is to be noted that although linearity is essential for the method proposed, functions of physical parameters can always be redefined as the primary parameters, in order to satisfy this assumption. For example, if J (inertia) and k (spring constant) are physical parameters and $\frac{k}{J}$ is a transfer function coefficient, we label p := $\frac{k}{J}$ as the new primary parameter of interest. Thus a broad class of practical problems can be effectively treated by this approach. In particular, the radius of the largest stability hypersphere, for a given controller, measures the stability margin provided by the controller, and can be used to compare the robustness of alternative designs.

2. PROBLEM FORMULATION

Consider the feedback system of Fig. 2.1 where G(s) is the plant transfer function and C(s) is the controller transfer function.

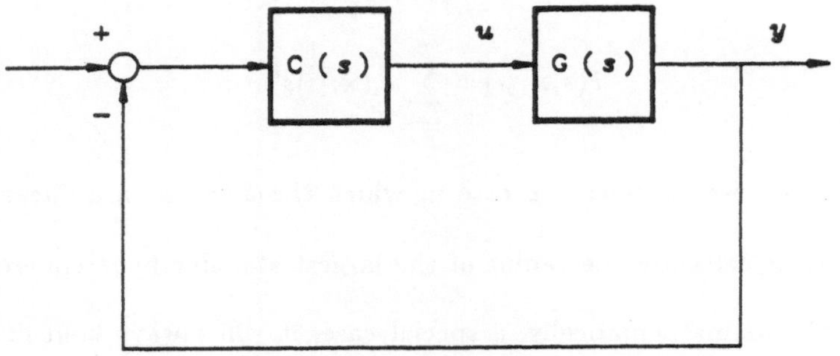

Figure 2.1.
Feedback System.

We shall develop our results specifically for single input (multioutput) or single output (multiinput) plants. Since the theory for the two cases are analogous we restrict our considerations to the single input case.

Let

$$\mathbf{G}(s) = \begin{pmatrix} \frac{\hat{n}_1(s)}{\hat{d}_1(s)} \\ \cdot \\ \cdot \\ \cdot \\ \frac{\hat{n}_m(s)}{\hat{d}_m(s)} \end{pmatrix} = \frac{1}{d(s)} \begin{pmatrix} n_1(s) \\ \cdot \\ \cdot \\ \cdot \\ n_m(s) \end{pmatrix} := \mathbf{n}(s)d^{-1}(s) \ , \tag{2.1}$$

where d(s) is the least common denominator of all elements of G(s) and write

$$d(s) = d_q s^q + \ \ldots\ldots + d_0 \ , \tag{2.2}$$

$$\mathbf{n}(s) = \mathbf{n}_q s^q + \ \ldots\ldots + \mathbf{n}_0 \ .$$

where d_i are real scalars, $\mathbf{n}_i$ are $m \times 1$ constant real vectors and $\mathbf{n}$ and d are right coprime.

The coefficients d_i and $\mathbf{n}_i$ of a physical system are generally subject to perturbation and therefore we introduce the nominal values denoted by d_i^0, $\mathbf{n}_i^0$ and the perturbations denoted by Δd_i, $\Delta \mathbf{n}_i$, $\ i = 0, 1, ..., q$. To rule out trivial cases, we will assume that the coprimeness assumed above continues to hold under the perturbations considered.

Let

$$\mathbf{p} := [\mathbf{n}_0^{\mathrm{T}}, d_0, \ \ldots, \mathbf{n}_q^{\mathrm{T}}, d_q]^{\mathrm{T}} \tag{2.3}$$

denote the plant parameter vector in R^k, $\ k = (1+m)(1+q)$,

$$\mathbf{p}^0 := [\mathbf{n}_0^{0\mathrm{T}}, d_0^0, \ \ldots, \mathbf{n}_q^{0\mathrm{T}}, d_q^0]^{\mathrm{T}} \tag{2.4}$$

its nominal value and

$$\Delta\mathbf{p} := [\Delta\mathbf{n}_0^{\mathrm{T}}, \Delta d_0, \ \ldots, \Delta\mathbf{n}_q^{\mathrm{T}}, \Delta d_q]^{\mathrm{T}} \tag{2.5}$$

its perturbation, so that

$$\mathbf{p} = \mathbf{p}^0 + \Delta\mathbf{p} \ . \tag{2.6}$$

The size of the real perturbation vector $\Delta\mathbf{p}$ is measured by its Euclidean length, denoted by $||\Delta\mathbf{p}||_2$ and is given by

$$||\Delta\mathbf{p}||_2^2 = ||\Delta\mathbf{n}_0||_2^2 + ||\Delta\mathbf{n}_1||_2^2 + \ldots + ||\Delta\mathbf{n}_q||_2^2 + (\Delta d_0)^2 + \ldots + (\Delta d_q)^2.$$

We shall assume that the controller $\mathbf{C}(s)$ stabilizes the nominal closed loop system when $\mathbf{p} = \mathbf{p}^0$. The problem to be solved in the next section is: Determine, for the given stabilizing controller $\mathbf{C}(s)$, the largest stability hypersphere $\mathbf{S}_\rho(\mathbf{p}^0)$, centered at $\mathbf{p}^0$, with radius $\rho(\mathbf{p}^0)$ so that the closed loop system is stable for all $\mathbf{p}\epsilon\mathbf{S}_\rho(\mathbf{p}^0)$, i.e. $\forall||\Delta\mathbf{p}||_2 < \rho(\mathbf{p}^0)$, and is unstable for at least one $\overline{\Delta\mathbf{p}}$ with $||\overline{\Delta\mathbf{p}}||_2 = \rho(\mathbf{p}^0)$.

3. THE STABILITY HYPERSPHERE: LINEAR CASE

Let

$$\mathbf{C}(s) = \frac{1}{d_c(s)}(n_{c1}(s), \ldots, n_{cm}(s)) := d_c^{-1}(s)\mathbf{n}_c^T(s) \tag{3.1}$$

and write

$$d_c(s) = d_{cp}(s)s^p + \ldots + d_{c0} \tag{3.2}$$

$$\mathbf{n}_c^T(s) = \mathbf{n}_{cp}^T s^p + \ldots + \mathbf{n}_{c0}^T$$

where d_{ci} are real scalars, $\mathbf{n}_{ci}$ are $1 \times m$ constant real vectors, and d_c, $\mathbf{n}_c^T$ are left coprime.

The characteristic polynomial of the closed loop system of Fig. 2.1 is given from (2.1) and (3.1) by the $p+q$ degree polynomial

$$\delta(s) = d_c(s)d(s) + \mathbf{n}_c^T(s)\mathbf{n}(s)\ . \tag{3.3}$$

Write

$$\delta(s) := \delta_0 + \delta_1 s + \ldots + \delta_n s^n\ ,\ n := p+q \tag{3.4}$$

and define the closed loop characteristic vector by

$$\delta := [\delta_n, \delta_{n-1}, \ldots, \delta_0]^T \in R^{n+1} . \tag{3.5}$$

Definition The characteristic vector $\delta \in R^{n+1}$ is said to be Hurwitz if and only if the corresponding polynomial $\delta(s)$ in (3.4) is strictly Hurwitz (i.e., has roots in the open left half plane). The set of Hurwitz vectors in R^{n+1} corresponding to n^{th} degree Hurwtz polynomials is denoted by $H_n \subset R^{n+1}$.

The controller $\mathbf{C}(s)$ may be viewed as an operator that maps the plant parameter vector $\mathbf{p}$ into the closed loop characteristic vector δ. Let δ^0 denote the image under this mapping of the nominal plant parameter $\mathbf{p}^0$, as defined in (2.4). If $\underline{\Delta} := \{\Delta\mathbf{p}\}$ denotes a given class of perturbation of $\mathbf{p}^0$ against which stabilization is required, the controller $\mathbf{C}(s)$ must map $\mathbf{p}^0 + \Delta\mathbf{p}$ into $\delta \in H_n$ for every $\Delta\mathbf{p} \in \underline{\Delta}$. These ideas can be sharpened by the following Lemma which makes the above mentioned mapping specific.

Lemma 3.1

The closed loop characteristic vector δ satisfies

$$\mathbf{X}\mathbf{p} = \delta \tag{3.6}$$

as well as

$$\mathbf{P}\mathbf{x} = \delta \tag{3.7}$$

where $\mathbf{X} \in R^{(p+q+1)\times[(1+m)(1+q)]}$, $\mathbf{P} \in R^{(p+q+1)\times[(1+m)(p+1)]}$, $\mathbf{x} \in R^{(1+m)(p+1)}$

and $\mathbf{p} \in R^{(1+m)(1+q)}$

$$X = \begin{pmatrix} & & & & & & \mathbf{n}_{cp}^T & d_{cp} \\ & & \mathbf{n}_{cp}^T & d_{cp} & \cdot & \cdot & \cdot & \cdot \\ \mathbf{n}_{cp}^T & d_{cp} & \cdot & \cdot & \cdot & \cdot & \cdot & \cdot \\ \cdot & \cdot & \cdot & \cdot & \cdot & \cdot & \cdot & \cdot \\ \cdot & \cdot & \cdot & \cdot & \cdot & \cdot & \mathbf{n}_{c0}^T & d_{c0} \\ \cdot & \cdot & \mathbf{n}_{c0}^T & d_{c0} & & & & \\ \mathbf{n}_{c0}^T & d_{c0} & & & & & & \end{pmatrix} \tag{3.8}$$

$$P = \begin{pmatrix} \mathbf{n}_q^T & d_q & & & & & & \\ \cdot & \cdot & \mathbf{n}_q^T & d_q & & & & \\ \cdot & \cdot & \cdot & \cdot & & & & \\ \cdot & \cdot & \cdot & \cdot & \cdot & \cdot & & \\ \cdot & \cdot & \cdot & \cdot & \cdot & \cdot & \mathbf{n}_q^T & d_q \\ \mathbf{n}_0^T & d_0 & \cdot & \cdot & \cdot & \cdot & \cdot & \cdot \\ & & \mathbf{n}_0^T & d_0 & \cdot & \cdot & \cdot & \cdot \\ & & & & & & \cdot & \cdot \\ & & & & & & \cdot & \cdot \\ & & & & & & \mathbf{n}_0^T & d_0 \end{pmatrix} \tag{3.9}$$

$$\mathbf{x} = (\, \mathbf{n}_{cp} \quad d_{cp} \quad \cdot \quad \cdot \quad \cdot \quad \cdot \quad \cdot \quad \mathbf{n}_{c0} \quad d_{c0}\,)^T . \tag{3.10}$$

The proof of this lemma is a simple calculation obtained by equating coefficients in (3.3) and is omitted. The lemma shows that the compensator maps the parameter $\mathbf{p}$ into the characteristic vector δ via the <u>linear</u> transformation X in (3.8). Analogously the plant maps the compensator parameter vector $\mathbf{x}$ into δ via the linear transformation P in (3.9). To avoid trivial cases, we will assume throughout that X has full rank.

Now, define the following sets in the space of the characteristic vector δ:

$$\Delta_0 := \{\delta | \delta \in R^{n+1} \ , \ \delta_0 = 0\} \tag{3.11}$$

$$\Delta_n := \{\delta | \delta \in R^{n+1} \ , \ \delta_n = 0\} \tag{3.12}$$

and for any real ω,

$$\Delta(\omega) := \{\delta | \delta \in R^{n+1} \ , \ \delta(s) = (s^2 + \omega^2) l(s) \ , \ l(s) \text{ arbitrary}\} . \tag{3.13}$$

The inverse image of each of these sets with respect to the compensator map X in the parameter space of **p** is defined next:

$$\Pi_0 := X^{-1}(\Delta_0) := \{\mathbf{p}|\mathbf{p} \in R^k \ , \ X\mathbf{p} \in \Delta_0\} \tag{3.14}$$

$$\Pi_n := X^{-1}(\Delta_n) := \{\mathbf{p}|\mathbf{p} \in R^k \ , \ X\mathbf{p} \in \Delta_n\} \tag{3.15}$$

$$\Pi(\omega) := X^{-1}(\Delta(\omega)) := \{\mathbf{p}|\mathbf{p} \in R^k \ , \ X\mathbf{p} \in \Delta(\omega)\} \ . \tag{3.16}$$

Note that Π_0, Π_n and $\Pi(\omega)$ are subspaces of R^k and Δ_0, Δ_n and $\Delta(\omega)$ are subspaces of R^{n+1}.

Let r_0, r_n and $r(\omega)$ denote the Euclidean distances between $\mathbf{p}^0$ and Π_0, Π_n and $\Pi(\omega)$ respectively. Then

$$r_0 := ||\mathbf{p}^0 - \mathbf{t}_0^*||_2 \tag{3.17}$$

where $\mathbf{t}_0^* \in \Pi_0$ and

$$||\mathbf{t}_0^* - \mathbf{p}^0||_2 \le ||\mathbf{t}_0 - \mathbf{p}^0||_2 \ , \ \forall \ \mathbf{t}_0 \in \Pi_0 \ . \tag{3.18}$$

Similarly

$$r_n := ||\mathbf{p}^0 - \mathbf{t}_n^*||_2 \tag{3.19}$$

where $\mathbf{t}_n^* \in \Pi_n$ and

$$||\mathbf{t}_n^* - \mathbf{p}^0||_2 \le ||\mathbf{t}_n - \mathbf{p}^0||_2 \ , \ \forall \ \mathbf{t}_n \in \Pi_n \tag{3.20}$$

and

$$r(\omega) := ||\mathbf{p}^0 - \mathbf{t}^*(\omega)||_2 \tag{3.21}$$

where $\mathbf{t}^*(\omega) \in \Pi(\omega)$ and

$$||\mathbf{t}^*(\omega) - \mathbf{p}^0||_2 \le ||\mathbf{t}(\omega) - \mathbf{p}^0||_2 \ , \ \forall \ \mathbf{t}(\omega) \in \Pi(\omega) \ . \tag{3.22}$$

Let

$$r := \inf_{0 \leq \omega < \infty} r(\omega) . \tag{3.23}$$

We are now ready to state one of the main results of this chapter.

<u>Theorem 3.1</u>

<u>Let C(s) be a fixed stabilizing controller as in (3.1) and (3.2). Then the radius of the largest stability hypersphere in the space of p, centered at $\mathbf{p}^0$, is given by,</u>

$$\rho(\mathbf{p}^0) = \min\{r_0,\ r_n\ , r\} . \tag{3.24}$$

The proof of this theorem is given in Section 5. Figure 3.1 illustrates this theorem. The intuitive content of the theorem is that a perturbation of $\Delta(\mathbf{p})$ of $\mathbf{p}^0$ cannot destabilize the closed loop system unless $\Delta\mathbf{p}$ intersects Π_0, Π_n or $\Pi(\omega)$ for some $\omega\epsilon[0,\ \infty]$, say ω^*. Intersection with Π_0 results in a root of $\delta(s)$ at the origin, intersection with Π_n results in a root at $s = \infty$, and intersection with $\Pi(\omega)$ results in complex conjugate roots at $s = \pm j\omega$, all of which are destabilizing.

An obvious extension of Theorem 3.1 will allow us to solve the robust stability problem for a general perturbation region. This is stated in Section 6.

Next we give formulas for the calculation of the distances r_0, r_n and r. Let $\delta \in \Delta_0$, $\mathbf{t} \in \Pi_0$ and let $\mathbf{w}_1 = [0, 0,\ \ldots, 0, 1]^T \in R^{n+1}$. Then from (3.6)

$$\delta = \mathrm{X}\mathbf{t} \tag{3.25}$$

and from (3.11),

$$\mathbf{w}_1^T\delta = \delta_0 = 0 = \mathbf{w}_1^T\mathrm{X}\mathbf{t} . \tag{3.26}$$

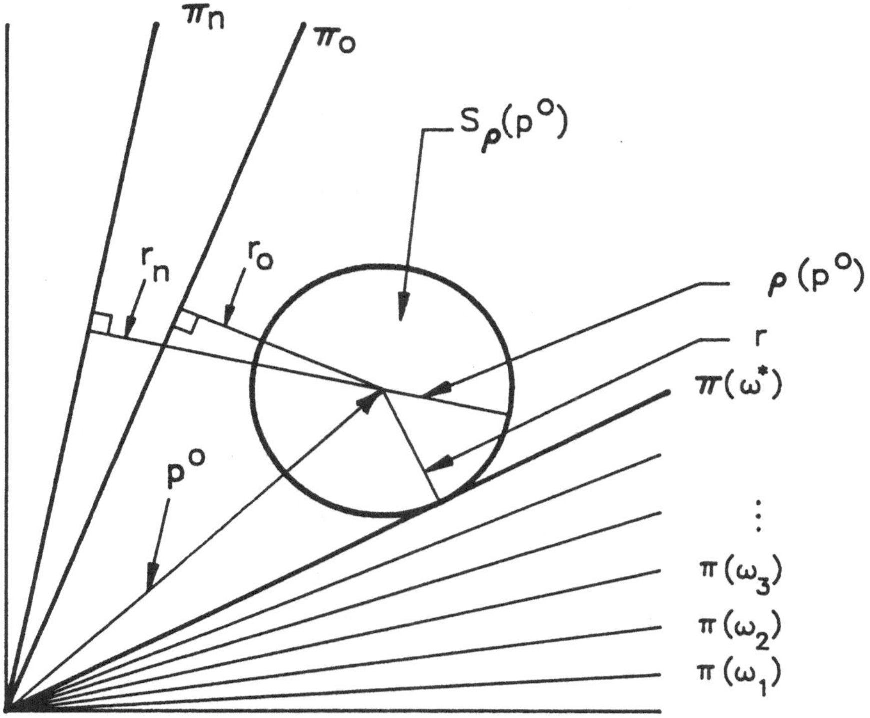

The Stability Hypersphere $S_\rho(p^o)$

Figure 3.1

Let $\mathbf{X}_l$ denote the last row of $\mathbf{X}$. Then (3.26) can be rewritten

$$\mathbf{X}_l \mathbf{t} = 0 \tag{3.27}$$

which shows that the shortest distance from $\mathbf{p}^0$ to Π_0 must lie on the normal $\mathbf{X}_l^T$ to Π_0. Therefore the shortest vector is given by

$$\mathbf{p}^0 - \mathbf{t} = \alpha \mathbf{X}_l^T . \tag{3.28}$$

To determine α, premultiply (3.28) by $\mathbf{X}_l$ and use (3.27)

$$\mathbf{X}_l \mathbf{p}^0 - \mathbf{X}_l \mathbf{t} = \alpha \mathbf{X}_l \mathbf{X}_l^T \tag{3.29}$$

$$\mathbf{X}_l \mathbf{p}^0 = \alpha \mathbf{X}_l \mathbf{X}_l^T \tag{3.30}$$

so that

$$\alpha = \frac{\mathbf{X}_l \mathbf{p}^0}{\mathbf{X}_l \mathbf{X}_l^T} . \tag{3.31}$$

Therefore the distance r_0 from $\mathbf{p}^0$ to the hyperplane Π_0 is given by

$$r_0^2 = \frac{1}{\mathbf{X}_l \mathbf{X}_l^T} [\mathbf{p}^{0T} \mathbf{X}_l^T \mathbf{X}_l \mathbf{p}^0] . \tag{3.32}$$

By exactly analogous arguments the distance r_n is calculated as

$$r_n^2 = \frac{1}{\mathbf{X}_f \mathbf{X}_f^T} [\mathbf{p}^{0T} \mathbf{X}_f^T \mathbf{X}_f \mathbf{p}^0] , \tag{3.33}$$

where $\mathbf{X}_f$ denotes the first row of $\mathbf{X}$.

The calculation of the distance $r(\omega)$ is now given. A representative vector $\delta \in \Delta(\omega)$ is given by

$$\delta = \Phi(\omega)\, \mathbf{1} \tag{3.34}$$

where

$$\Phi(\omega) = \begin{pmatrix} 1 & & & \\ 0 & 1 & & \\ \omega^2 & 0 & \cdot & \\ & \omega^2 & \cdot & 1 \\ & & \cdot & 0 \\ & & & \omega^2 \end{pmatrix} \in R^{(n+1)\times(n-1)} \tag{3.35}$$

and $\mathbf{l} \in R^{n-1}$ is arbitrary.

If $\mathbf{t}(\omega) \in \Pi(\omega)$ then

$$\mathrm{X}\mathbf{t}(\omega) = \Phi(\omega)\mathbf{l}\ . \tag{3.36}$$

The matrix X is of size $(p+q+1) \times [(m+1)(1+q)]$. Consider first the case where X has at least as many columns as rows. When the plant order q is greater than or equal to the controller order p, X will have full row rank.

Case I : $m + mq \geq p$

We partition X and $\mathbf{t}(\omega)$ as follows

$$\mathrm{X} = [\mathrm{X}_I,\ \mathrm{X}_J]$$

$$\mathbf{t}(\omega) = \begin{bmatrix} \mathbf{t}_I(\omega) \\ \mathbf{t}_J(\omega) \end{bmatrix}$$

where X_I is square and nonsingular. Such a partition can be made without loss of generality. This is proved in Section 5 where a specific way of constructing the nonunique matrices X_I and X_J is also given. Now,

$$\mathrm{X}_I\mathbf{t}_I(\omega) = \Phi(\omega)\mathbf{l} - \mathrm{X}_J\mathbf{t}_J(\omega) \tag{3.37}$$

so that

$$\mathbf{t}_I(\omega) = \mathrm{X}_I^{-1}\Phi(\omega)\mathbf{l} - \mathrm{X}_I^{-1}\mathrm{X}_J\mathbf{t}_J(\omega).$$

Then a representative vector $\mathbf{t}(\omega) \in \Pi(\omega)$ is given by,

$$\mathbf{t}(\omega) = \underbrace{\begin{pmatrix} \mathbf{X}_I^{-1}\Phi(\omega) & -\mathbf{X}_I^{-1}\mathbf{X}_J \\ 0 & 1 \end{pmatrix}}_{\mathbf{P}(\omega)} \underbrace{\begin{pmatrix} 1 \\ \mathbf{t}_J \end{pmatrix}}_{\mathbf{l}_t} \tag{3.38}$$

$$:= \mathbf{P}(\omega)\, \mathbf{l}_t \tag{3.39}$$

where $\mathbf{P}(\omega)$ is a fixed real matrix for each ω and $\mathbf{l}_t$ is an arbitrary real vector. Note that the dependence of $\mathbf{t}_J$ on ω can be dropped since $\mathbf{t}_J$ can be any vector. By letting $\mathbf{l}_t$ sweep over all real vectors in (3.39) we generate all solutions of (3.36). Now

$$\mathbf{t}(\omega) - \mathbf{p}^0 = \mathbf{P}(\omega)\mathbf{l}_t - \mathbf{p}^0 \tag{3.40}$$

and

$$||\mathbf{t}(\omega) - \mathbf{p}^0||_2^2 = \mathbf{p}^{0^{\mathrm{T}}}\mathbf{p}^0 - 2\mathbf{l}_t^{\mathrm{T}}\mathbf{P}^{\mathrm{T}}(\omega)\mathbf{p}^0 + \mathbf{l}_t^{\mathrm{T}}\mathbf{P}^{\mathrm{T}}(\omega)\mathbf{P}(\omega)\mathbf{l}_t \ . \tag{3.41}$$

The vector $\mathbf{l}_t = \mathbf{l}_t^*$ which minimizes the distance (3.41) for a fixed ω is given by setting the corresponding gradient to zero. This gives

$$\mathbf{l}_t^* = (\mathbf{P}^{\mathrm{T}}(\omega)\mathbf{P}(\omega))^{-1}\mathbf{P}^{\mathrm{T}}(\omega)\mathbf{p}^0 \tag{3.42}$$

and

$$r^2(\omega) = \mathbf{p}^{0^{\mathrm{T}}} \underbrace{(\mathbf{I} - \mathbf{P}(\omega)(\mathbf{P}^{\mathrm{T}}(\omega)\mathbf{P}(\omega))^{-1}\mathbf{P}^{\mathrm{T}}(\omega))}_{\mathbf{Q}(\omega)} \mathbf{p}^0 \ . \tag{3.43}$$

<u>Case II : $p > m + mq$</u>

In this case the equation (3.36) may or may not have a solution for a fixed ω and $\mathbf{l}$. If a solution exists, it is unique and is given by

$$\mathbf{t}(\omega) = \underbrace{(\mathbf{X}^{\mathrm{T}}\mathbf{X})^{-1}\mathbf{X}^{\mathrm{T}}\Phi(\omega)}_{\overline{\mathbf{P}}(\omega)}\mathbf{l} \tag{3.44}$$

where l satisfies

$$[X(X^TX)^{-1}X^T - I]\Phi(\omega)l = [X\overline{P}(\omega) - \Phi(\omega)]l = 0 \,. \tag{3.45}$$

Let

$$l = N(\omega)l_t \tag{3.46}$$

denote the solutions of (3.45) with l_t arbitrary so that

$$t(\omega) = \underbrace{\overline{P}(\omega)N(\omega)}_{P(\omega)} l_t := P(\omega)l_t \,. \tag{3.47}$$

Now (3.47) gives the solution set of (3.36) as l_t ranges over all real vectors. Since (3.47) and (3.39) are of the same form we can apply the preceding development (3.39)–(3.42) to obtain the expression (3.43) for $r^2(\omega)$.

Now, in both cases we have

$$r^2 = \inf_\omega r^2(\omega) = \inf_\omega p^{0^T}Q(\omega)p^0. \tag{3.48}$$

Because $\Phi(\omega)$ and hence $P(\omega)$ and $Q(\omega)$ are known, the function $r^2(\omega)$ can be determined and the minimization in (3.48) can be carried out numerically (for instance by graphing $r^2(\omega)$) and the global minimum in (3.48) can be found over $0 \le \omega < \infty$. Since $r^2(\omega)$ is a continuous function of ω^2 the minimum of $r^2(\omega)$ occurs either for a finite value of ω or at $\omega = \infty$. The latter case corresponds to $\delta_n = 0$, $\delta_{n-1} = 0$ (i.e. a double pole at infinity) and since $\Delta(\infty) \subset \Delta_n$, $r(\infty) \ge r_n$. Similarly, since $\Delta(0) \subset \Delta_0$, if the minimum of $r(\omega)$ occurs at $\omega = 0$, we have $r(0) \ge r_0$. Therefore, the global minimum of $r(\omega)$ need be found only in the interior of the interval $0 \le \omega \le \infty$.

Before concluding this section, we point out the interesting fact that the stability hypersphere determined here can be translated linearly to determine a larger conical stability region. Let $\mathbf{S}_\rho(\mathbf{p}^0)$ be the stability hypersphere determined by Theorem 3.1, i.e.

$$\mathbf{S}_\rho(\mathbf{p}^0) := \{\mathbf{p} | \mathbf{p} \epsilon R^k, \ \mathbf{p} = \mathbf{p}_0 + \Delta\mathbf{p}, \|\Delta p\|_2 < \rho(\mathbf{p}_0)\} \tag{3.49}$$

and introduce the cone

$$\mathbf{C}_\rho(\mathbf{p}^0) = \{\mathbf{t} | \mathbf{t} \epsilon R^k, \ \mathbf{t} = \alpha\mathbf{p}, \ \mathbf{p} \epsilon \mathbf{S}_\rho(\mathbf{p}^0), 0 < \alpha < \infty\}. \tag{3.50}$$

Corollary 3.2

Under the conditions of Theorem 3.1, the conical region $\mathbf{C}_\rho(\mathbf{p}^0)$ is a stability region.

The proof of this result simply depends on the fact that if $\mathbf{p}$ is "stable" so is $\alpha\mathbf{p}$ for $0 < \alpha < \infty$ because of the linear equation (3.6). Therefore, $X\mathbf{S}_\rho(\mathbf{p}^0) \subset H_n$ implies that $X\mathbf{C}_\rho(\mathbf{p}^0) \subset H_n$. This corollary is illustrated in Figure 3.2.

The above calculations show that Theorem 3.1 provides a constructive procedure for calculating the stability hypersphere. In the next section we show how this calculation can be extended to handle situations where the transfer function coefficients are interdependent.

4. THE STABILITY HYPERSPHERE: AFFINE CASE

Let

$$\mathbf{a} := [a_1, \ a_2, \ \ldots, \ a_l]^T \tag{4.1}$$

denote the vector of primary parameters in R^l,

$$\mathbf{a}^0 := [a_1^0, \ a_2^0, \ \ldots, \ a_l^0]^T \tag{4.2}$$

its nominal value and

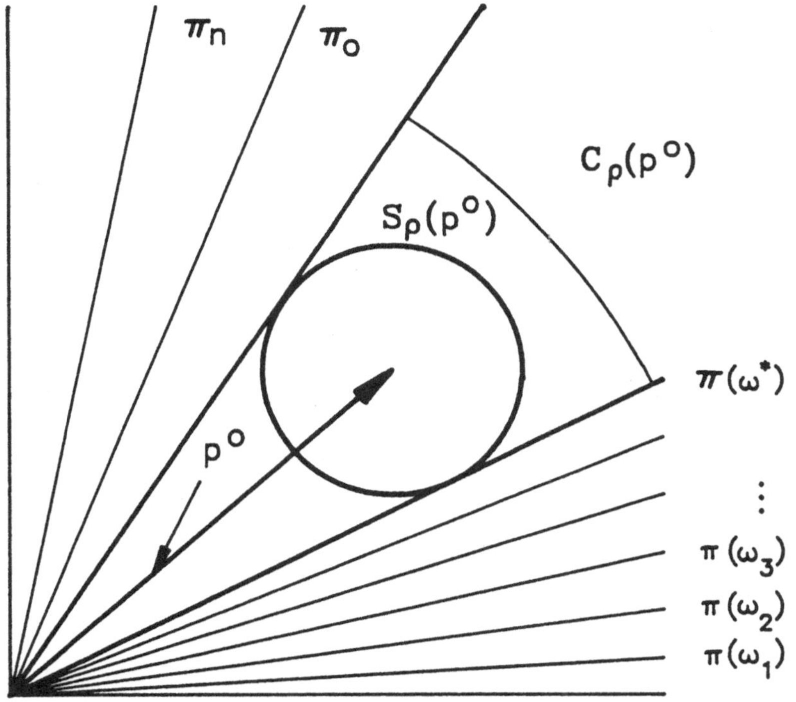

The Stability Hypercone $C_\rho(p^o)$

Figure 3.2

$$\Delta\mathbf{a} := [\Delta a_1,\ \Delta a_2,\ \ldots.,\ \Delta a_l]^{\mathrm{T}} \tag{4.3}$$

its perturbation, so that

$$\mathbf{a} = \mathbf{a}^0 + \Delta\mathbf{a}\ . \tag{4.4}$$

Let us assume that the vector $\mathbf{p}$ in (2.3) of the plant transfer function coefficients depends affinely on $\mathbf{a}$ as

$$\mathbf{p} = \mathrm{A}\mathbf{a} + \mathbf{b}\ , \tag{4.5}$$

where $\mathrm{A} \in R^{k \times l}$ and $\mathbf{b} \in R^k$. Without loss of generality, we can assume that A is of full column rank and therefore $l \leq k$ (otherwise the parameters $\mathbf{a}$ could be redefined).

According to (3.6) and (4.5) the closed loop characteristic polynomial vector δ is now expressed as

$$\mathrm{X}\mathrm{A}\mathbf{a} + \mathrm{X}\mathbf{b} = \delta \tag{4.6}$$

which shows the affine transformation mapping the parameter vector $\mathbf{a}$ into the characteristic vector δ. As before, let us consider the sets (3.11) - (3.13) and denote the inverse images of Δ_0, Δ_n and $\Delta(\omega)$ (analogous to (3.14) - (3.16)) in the space of $\mathbf{a}$ as Π_0, Π_n and $\Pi(\omega)$. It is to be noted, however, that now some of these sets may be empty. Therefore, definitions (3.17), (3.19) and (3.21) with $\mathbf{a}$ substituted for $\mathbf{p}$ have to be augmented by

$$r_0 = \infty \ \text{ if } \ \Pi_0 = \emptyset \tag{4.7}$$

$$r_n = \infty \ \text{ if } \ \Pi_n = \emptyset \tag{4.8}$$

$$r(\omega) = \infty \ \text{ if } \ \Pi(\omega) = \emptyset\ . \tag{4.9}$$

Defining r as in (3.23) we can generalize Theorem 3.1 as follows:

Theorem 4.1

Let C(s) be a fixed stabilizing controller as in (3.1) and (3.2). Then the radius of the largest stability hypersphere in the space of primary parameters centered at $\mathbf{a}^0$, is given by

$$\rho(\mathbf{a}^0) = \min\{r_0, r_n, r\} . \tag{4.10}$$

The proof of this theorem is similar to the proof of Theorem 3.1 and is omitted.

We now give formulas for the calculation of the distances r_0, r_n and r in the space of $\mathbf{a}$. We note that $\mathbf{t} \in \Pi_0$ if and only if

$$X_l A\mathbf{t} + X_l \mathbf{b} = \mathbf{0} . \tag{4.11}$$

The above equation fails to hold if and only if the vector $X_l A = \mathbf{0}$, $X_l \mathbf{b} \neq 0$ and then Π_0 is empty ($r_0 = \infty$). Otherwise, if $X_l A \neq \mathbf{0}$, the distance r_0 of the point $\mathbf{a}^0$ from the hyperplane (4.11) is given by the formula

$$r_0 = \frac{1}{||X_l A||_2} |X_l A\mathbf{a}^0 + X_l \mathbf{b}| \tag{4.12}$$

or

$$r_0^2 = \frac{1}{X_l A A^T X_l^T} \left[\mathbf{a}^{0^T} A^T X_l A \mathbf{a}^0 + 2\mathbf{a}^{0^T} A^T X_l^T X_l \mathbf{b} + \mathbf{b}^T X_l^T X_l \mathbf{b}\right]. \tag{4.13}$$

The distance r_n is calculated similarly as

$$r_n = \begin{cases} \infty & \text{if } X_f A = \mathbf{0} \text{ and } X_f \mathbf{b} \neq 0; \\ \frac{1}{||X_f A||_2} |X_f A\mathbf{a}^0 + X_f \mathbf{b}|, & \text{if } X_f A \neq \mathbf{0} \end{cases} \tag{4.14}$$

It should be mentioned that it is not possible that $X_l A = \mathbf{0}$ and $X_l \mathbf{b} = 0$ (or $X_f A = \mathbf{0}$ and $X_f \mathbf{b} = 0$) simultaneously, if the nominal point $\mathbf{a}^0$ is stabilized by C(s).

The calculation of the distance $r(\omega)$ is now given. After exactly analogous derivations as in (3.34) - (3.38) the formula (3.38) can be written as

$$\mathbf{t}(\omega) = \mathrm{A}\mathbf{t_a}(\omega) + \mathbf{b} = \mathrm{P}(\omega)\mathbf{l}_t \tag{4.15}$$

where now $\mathbf{t_a}(\omega) \in \Pi(\omega)$ in the space of **a**.

Since A is of full column rank it can be, after some possible row interchanges, partitioned as

$$\mathrm{A} = \begin{pmatrix} \mathrm{A}_1 \\ \mathrm{A}_2 \end{pmatrix} \tag{4.16}$$

where A_1 is a square nonsingular matrix, $\mathrm{A}_1 \in R^{l\times l}$, $\mathrm{A}_2 \in R^{(k-l)\times l}$. Equation (4.15), after the same row interchanges, can be expressed as

$$\begin{pmatrix} \mathrm{A}_1 \\ \mathrm{A}_2 \end{pmatrix} \mathbf{t_a}(\omega) = \begin{pmatrix} \mathrm{P}_1(\omega) \\ \mathrm{P}_2(\omega) \end{pmatrix} \mathbf{l}_t - \begin{pmatrix} \mathbf{b}_1 \\ \mathbf{b}_2 \end{pmatrix} . \tag{4.17}$$

From the first part of (4.17) we get

$$\mathbf{t_a}(\omega) = \mathrm{A}_1^{-1}\mathrm{P}_1(\omega)\mathbf{l}_t - \mathrm{A}_1^{-1}\mathbf{b}_1 \tag{4.18}$$

which substituted into the second part of (4.17) gives

$$\underbrace{[\mathrm{A}_2\mathrm{A}_1{}^{-1}\mathrm{P}_1(\omega) - \mathrm{P}_2(\omega)]\,\mathbf{l}_t}_{\mathrm{B}(\omega)} = \underbrace{\mathrm{A}_2\mathrm{A}_1{}^{-1}\mathbf{b}_1 - \mathbf{b}_2}_{\mathbf{c}} \tag{4.19}$$

or

$$\mathrm{B}(\omega)\mathbf{l}_t = \mathbf{c} \ . \tag{4.20}$$

Equation (4.20) is of primary importance in determining whether $\Pi(\omega)$ is empty or not. Let

$$\Omega := \{\omega |\ \mathrm{rank}\ [\mathrm{B}(\omega)] = \mathrm{rank}\ [\mathrm{B}(\omega)|\mathbf{c}]\ ,\ 0 \le \omega < \infty\}. \tag{4.21}$$

Then $\Pi(\omega) \neq \emptyset$ if and only if $\omega \in \Omega \neq \emptyset$ and therefore we take

$$r(\omega) = \infty \quad \text{for } \omega \notin \Omega. \tag{4.22}$$

For $\omega \in \Omega$ (if $\Omega \neq \emptyset$) the general solution of (4.20) is of the form

$$\mathbf{l}_t = \mathrm{D}(\omega)\bar{\mathbf{l}}_t + \mathbf{e}(\omega) \tag{4.23}$$

and therefore

$$\mathbf{t_a}(\omega) = \tilde{P}(\omega)\bar{\mathbf{l}}_t + \tau(\omega) \tag{4.24}$$

is the general solution of (4.15) with $\bar{\mathbf{l}}_t$ an arbitrary real vector. Now, in a manner analogous to (3.40) - (3.43), we obtain

$$r^2(\omega) = (\mathbf{a}^0 - \tau(\omega))^{\mathrm{T}}\tilde{\mathrm{Q}}(\omega)(\mathbf{a}^0 - \tau(\omega)) \tag{4.25}$$

and finally

$$r^2 = \inf_{\omega\in\Omega} r^2(\omega)\ . \tag{4.26}$$

The above derivation allows us to effectively deal with a more general class of perturbations than that covered in Section 3, i.e., some plant transfer function coefficient can be interdependent and some can be fixed.

It is worth mentioning that the way we have dealt with equations (4.6) and (3.34) in this section gives a good theoretical insight into the problem. However, in a specific example it may be simpler to solve (4.6) directly with the right hand side set equal to (3.34).

As a final remark, we note that Theorem 1.3.2 of Chapter 1 is a special case of Theorem 2.3.1 of the last section obtained by setting $\mathrm{X} = I_{n+1}$, $\mathrm{p} = \delta$.

5. PROOF OF THE MAIN RESULT

This section contains the proof of Theorem 3.1, and the demonstration of nonsingularity of X_I in (3.36).

Proof of Theorem 3.1

Let $C = C^+ \cup C^-$ denote the complex plane with

$$C^+ = \{s|\text{Re } s \geq 0\} \ , \ C^- = \{s|\text{Re } s < 0\}$$

and let $C_I \in C^+$ denote the imaginary axis. Let $Z(\delta)$ denote the zeros of the polynomial $\delta(s) = \delta_0 + \delta_1 s + \ .\ .\ .\ .\ \delta_n s^n$ and introduce the function

$$\delta(s) \rightarrow [\delta_n, \delta_{n-1}, \ .\ .\ .\ .\ , \delta_0]^T := \delta \in R^{n+1} \ .$$

As in Section 3 let

$$\Delta_0 := \{\delta | 0 \in Z(\delta)\} \tag{5.1}$$

$$\Delta_n := \{\delta | \delta_n = 0\} \tag{5.2}$$

and for $0 \leq \omega < \infty$

$$\Delta(\omega) := \{\delta | \delta(s) = (s^2 + \omega^2) l(s), \ l(s) \text{ arbitrary}\}. \tag{5.3}$$

Define

$$\Delta_I := \{\delta | Z(\delta) \cap C_I \neq \emptyset\} \tag{5.4}$$

$$\Delta^- := \{\delta | Z(\delta) \subset C^-\} \tag{5.5}$$

$$\Delta^+ := \{\delta | Z(\delta) \cap C^+ \neq \emptyset\} \tag{5.6}$$

and let H_n denote the set of nth degree polynomials with zeroes in C^-:

$$H_n := \{\delta | \delta \in R^{n+1},\ \delta_n \neq 0,\ Z(\delta) \in C^-\} \tag{5.7}$$

or

$$H_n = \Delta^- \setminus \Delta_n. \tag{5.8}$$

We also note that

$$\Delta_I = \bigcup_{0 \leq \omega < \infty} (\Delta(\omega)) \cup \Delta_0. \tag{5.9}$$

Now consider the closed loop system of Fig. 2.1 and the equation

$$\mathbf{X}\mathbf{p} = \delta$$

for the characteristic vector. With the compensator, i.e. X, fixed and the plant parameter $\mathbf{p} = \mathbf{p}^0 + \Delta\mathbf{p}$ we have $\delta = \delta(\mathbf{x}, \mathbf{p}^0 + \Delta\mathbf{p})$ and closed loop stability is equivalent to

$$\delta(\mathbf{x}, \mathbf{p}^0 + \Delta\mathbf{p}) \in H_n. \tag{5.10}$$

Let

$$\rho(\mathbf{p}^0) = \min\{r_0, r_n, r\} \tag{5.11}$$

as in (3.24) and let $\mathbf{S}_\rho(\mathbf{p}^0)$ denote the interior of the hypersphere of radius $\rho(\mathbf{p}^0)$ centered at $\mathbf{p}^0$ in parameter space:

$$\mathbf{S}_\rho(\mathbf{p}^0) := \{\mathbf{p} | \mathbf{p} \in R^k,\ \mathbf{p} = \mathbf{p}^0 + \Delta\mathbf{p},\ ||\Delta\mathbf{p}||_2 < \rho(\mathbf{p}^0)\}. \tag{5.12}$$

Let $\mathbf{S}_\rho^B(\mathbf{p}^0)$ denote the boundary of this hypersphere:

$$\mathbf{S}_\rho^B(\mathbf{p}^0) := \{\mathbf{p} | \mathbf{p} \in R^k,\ \mathbf{p} = \mathbf{p}^0 + \Delta\mathbf{p},\ ||\Delta\mathbf{p}||_2 = \rho(\mathbf{p}^0)\}. \tag{5.13}$$

The proof of the theorem now consists of showing that

$$\delta(\mathbf{x},\mathbf{p}) \in H_n \ \ \forall \mathbf{p} \in \mathbf{S}_\rho(\mathbf{p}^0) \tag{5.14}$$

and

$$\delta(\mathbf{x},\mathbf{p}^*) \notin H_n \text{ for some } \mathbf{p}^* \in \mathbf{S}_\rho^B(\mathbf{p}^0) \tag{5.15}$$

which together show that $\mathbf{S}_\rho(\mathbf{p}^0)$ is the largest stability hypersphere. Note that $\mathbf{S}_\rho(\mathbf{p}^0)$ cannot intersect Π_n as otherwise $\delta(s)$ has a root at $s = \infty$ which causes instability of the closed loop system.

For clarity of presentation let us assume that

$$\rho = \rho(\mathbf{p}^0) = r_0. \tag{5.16}$$

Then there exists $\mathbf{t}_0^* \in \Pi_0$ (see 3.14) such that

$$r_0 = \|\mathbf{t}_0^* - \mathbf{p}^0\| \leq \|\mathbf{t}_0 - \mathbf{p}^0\|_2 \ \ \forall \mathbf{t}_0 \in \Pi_0. \tag{5.17}$$

Then

$$\mathbf{t}_0^* \in \mathbf{S}_{r_0}^B(\mathbf{p}^0) \tag{5.18}$$

and

$$\mathbf{X}\mathbf{t}_0^* := \delta_0^* \in \Delta_0 \tag{5.19}$$

and therefore

$$\mathbf{X}\mathbf{t}_0^* \notin H_n. \tag{5.20}$$

With $\mathbf{p}^* := \mathbf{t}_0^*$ and $\delta(\mathbf{x},\mathbf{p}^*) = \delta_0^*$ this proves (5.15), which shows that at least one point on the boundary of $\mathbf{S}_{r_0}(\mathbf{p}^0)$ corresponds to an unstable system.

To prove (5.14) we note that

$$||\mathbf{p}-\mathbf{p}^0||_2 < r_0 \leq \min\{r_n,\ r\} \quad \forall \mathbf{p} \in \mathbf{S}_{r_0}(\mathbf{p}^0). \tag{5.21}$$

Define

$$\Pi := \bigcup_{0\leq\omega<\infty} \Pi(\omega). \tag{5.22}$$

Now (5.21) implies that

$$\mathbf{S}_{r_0}(\mathbf{p}^0)\cap\Pi_0 = \emptyset \tag{5.23}$$

$$\mathbf{S}_{r_0}(\mathbf{p}^0)\cap\Pi_n = \emptyset \tag{5.24}$$

and

$$\mathbf{S}_{r_0}(\mathbf{p}^0)\cap\Pi = \emptyset. \tag{5.25}$$

Let

$$\mathrm{X}\mathbf{S}_{r_0}(\mathbf{p}^0) \ := \ \{\delta|\delta = \mathrm{X}\mathbf{p},\ \mathbf{p}\in\mathbf{S}_{r_0}(\mathbf{p}^0)\}. \tag{5.26}$$

Then (5.23) - (5.25) and the definition (3.14) - (3.16) of Π_0, Π_n and $\Pi(\omega)$ imply that

$$\mathrm{X}\mathbf{S}_{r_0}(\mathbf{p}^0)\cap\Delta_0 = \emptyset \tag{5.27}$$

$$\mathrm{X}\mathbf{S}_{r_0}(\mathbf{p}^0)\cap\Delta_n = \emptyset \tag{5.28}$$

and

$$\mathrm{X}\mathbf{S}_{r_0}(\mathbf{p}^0)\cap\Delta_I = \emptyset. \tag{5.29}$$

Now, since the function $Z(\delta)$ is continuous, the following well known result will hold.

Fact 1 If $\Delta \subset R^{n+1}$ is a simply connected region then $\Delta \subset H_n$ if and only if there exists $\delta^* \in \Delta$ such that $\delta^* \in H_n$ and $\Delta \cap H_n^B = \emptyset$, where H_n^B is the boundary of H_n.

From the definition of H_n

$$H_n^B \subset \Delta_I \cup \Delta_n = \Delta_0 \cup \Delta_n \cup \Delta_I \tag{5.30}$$

and (5.27) - (5.30) imply that

$$\mathbf{X}\mathbf{S}_{r_0}(\mathbf{p}^0) \cap H_n^B = \emptyset. \tag{5.31}$$

Since C(s) stabilizes, by assumption, the nominal system we have

$$\mathbf{X}\mathbf{p}^0 \in H_n. \tag{5.32}$$

We also note that $\mathbf{X}\mathbf{S}_{r_0}(\mathbf{p}^0)$ is a simply connected region in R^{n+1} since $\mathbf{S}_{r_0}(\mathbf{p}^0)$ is simply connected and δ is a continuous function of $\mathbf{p}$ and the number of roots of $\delta(\mathbf{s})$ does not change for $\mathbf{p} \epsilon\ \mathbf{S}_{r_0}(\mathbf{p}^0)$. Therefore, the conditions (5.31), (5.32) and Fact 1 imply that

$$\mathbf{X}\mathbf{S}_{r_0}(\mathbf{p}^0) \subset H_n\ . \tag{5.33}$$

This shows that all parameter points $\mathbf{p}$ inside the open hypersphere $\mathbf{S}_{r_0}(\mathbf{p}^0)$ of radius r_0 centered at $\mathbf{p}^0$ result in stable closed loop systems and completes the proof for the case $\rho = r_0$. When $\rho = r_n$ or $\rho = r$ exactly analogous arguments apply. These details are omitted. ◇

Construction of X$_I$ and Proof of its Invertibility

In this section we consider the equations (3.6) and (3.36) and show how to construct the nonsingular matrix X$_I$ in (3.36). In (3.36)

$$\mathrm{X} = \begin{pmatrix} & & & & & & & & \mathbf{n}_{cp} & d_{cp} \\ & & & & & & \mathbf{n}_{cp} & d_{cp} & \cdot & \cdot \\ & & \mathbf{n}_{cp} & d_{cp} & \cdot & \cdot & \cdot & \cdot & & \\ \mathbf{n}_{cp} & d_{cp} & \cdot & \cdot & \cdot & \cdot & \cdot & \cdot & \mathbf{n}_{c0} & d_{c0} \\ \cdot & \cdot & \cdot & \cdot & \cdot & \cdot & \cdot & \cdot & & \\ \cdot & \cdot & \cdot & \cdot & \cdot & \cdot & \mathbf{n}_{c0} & d_{c0} & & \\ \cdot & \cdot & \mathbf{n}_{c0} & d_{c0} & & & & & & \\ \mathbf{n}_{c0} & d_{c0} & & & & & & & & \end{pmatrix}$$

and

$$\mathbf{p}^{\mathrm{T}} = (\mathbf{n}_0^{\mathrm{T}} \quad , \quad d_0 \quad . \quad . \quad . \quad . \quad . \quad \mathbf{n}_q^{\mathrm{T}} \quad , \quad d_q)$$

where $\mathrm{X} \in R^{(q+p+1)\times[(1+m)(q+1)]}$ and $\mathbf{p} \in R^{[(1+m)(q+1)]}$.

Consider first the case m=1. In this case we may define

$$\mathrm{X}_I = \begin{pmatrix}
 & & & & & & & & & & d_{cp} \\
 & & & & & & & & & & . \\
 & & & & & & & & d_{cp} & . \quad . & . \\
 & & & & & & & & . & . \quad . & . \\
 & & & & & & n_{cp} & d_{cp} & . & . \quad . & d_{c0} \\
 & & n_{cp} & d_{cp} & . & . & . & . & . & . \quad . & \\
n_{cp} & d_{cp} & . & . & . & . & . & . & d_{c0} & & \\
. & . & . & . & . & . & . & . & & & \\
. & . & . & . & . & . & n_{c0} & d_{c0} & & & \\
. & . & n_{c0} & d_{c0} & & & & & & & \\
n_{c0} & d_{c0} & & & & & & & & &
\end{pmatrix} \in R^{(q+p+1)\times(q+p+1)}$$

It is easily shown using the eliminant matrix that X_I is nonsingular if $n_c(s)$, $d_c(s)$ are coprime as assumed.

For the general case we let $n_{cj}(s)$ be coprime with $d_c(s)$ for some $1 \le j \le m$. Then the matrix

$$\mathrm{X}_j = \begin{pmatrix}
 & & & & & & n_{cp}^j & d_{cp} \\
 & & n_{cp}^j & d_{cp} & . & . & . & . \\
n_{cp}^j & d_{cp} & . & . & . & . & . & . \\
. & . & . & . & . & . & . & . \\
. & . & . & . & . & . & n_{c0}^j & d_{c0} \\
. & . & n_{c0}^j & d_{c0} & & & & \\
n_{c0}^j & d_{c0} & & & & & &
\end{pmatrix} \in R^{2p\times 2p}$$

is nonsingular as before. Now let

$$
\mathrm{X}_I = \begin{pmatrix}
 & & & & & d_{cp} \\
 & & & & \cdot & \vdots \\
 & & & \cdot & & \vdots \\
 & & & d_{cp} & & \vdots \\
 & & \mathrm{X}_j & \cdot & & d_{c0} \\
 & & \cdot & \vdots & & \\
 & \cdot & & \vdots & & \\
 \cdot & & & d_{c0} & & \\
\mathrm{X}_j & & & & & \\
\mathrm{X}_j & & & & &
\end{pmatrix} \in R^{(q+p+1)\times(q+p+1)}
$$

Clearly X_I is nonsingular because X_j is nonsingular and $d_{cp} \neq 0$. Now from the form of X we can write

$$
\mathrm{X}\ \mathbf{p} = \mathrm{X}_I \mathbf{p}_I + \mathrm{X}_J \mathbf{p}_J
$$

by permuting the components of $\mathbf{p}$ to form $\mathbf{p}_I$ and $\mathbf{p}_J$. $\diamondsuit$

6. SOLUTION OF THE ROBUST STABILITY PROBLEM

The theory developed in sections 3 and 4 was concerned with determining the largest hypersphere centered at the nominal parameter that would retain stability. Theorems 3.1 and 4.1 show that this can be done by fitting the largest such hypersphere that just touches the closest of the regions Π_0, Π_n or Π. From the proof of Theorem 3.1 it is clear that this result in fact holds for any simply connected perturbation region containing the nominal parameter. This observation allows us to give a general solution to the robust stability problem. To state this, let $\mathbf{p}^0 \epsilon R^k$ in Theorem 3.1 or $\mathbf{a}^0 \epsilon R^\ell$ in Theorem 4.1 be the nominal parameter, let X correspond as in (3.6) to a nominally stabilizing controller, and let the class of perturbations to be handled, $\underline{\Delta}^k := \{\Delta\mathbf{p}\}$, or $\underline{\Delta}^\ell := \{\Delta\mathbf{a}\}$, be given with $\underline{\Delta}^k \subset R^k$, $\underline{\Delta}^\ell \subset R^\ell$ being arbitrary but <u>simply connected.</u> Let

$$
\mathcal{P} := \mathbf{p}^0 + \underline{\Delta}^k := \{\mathbf{p} | \mathbf{p} = \mathbf{p}^0 + \Delta\mathbf{p},\ \Delta\mathbf{p} \epsilon \underline{\Delta}^k\} \tag{6.1}
$$

and

$$\mathcal{A} := \mathbf{a}^0 + \underline{\Delta}^\ell := \{\mathbf{a} | \mathbf{a} = \mathbf{a}^0 + \Delta\mathbf{a}, \ \Delta\mathbf{a} \epsilon \underline{\Delta}^\ell\} \tag{6.2}$$

and let Π_0, Π_n and $\Pi(\omega)$ be defined as in Section 3 and Section 4.

Theorem 6.1

Let C(s) be a fixed stabilizing controller as in Theorem 3.1 or Theorem 4.1 of this chapter, which stabilizes the closed loop system for the nominal parameter $\mathbf{p} = \mathbf{p}^0$ or $\mathbf{a} = \mathbf{a}^0$. Then closed loop stability holds for all parameters contained in the simply connected region $\mathcal{P}$ if and only if

$$\mathcal{P} \bigcap \Pi_0 = \phi \tag{6.3a}$$

$$\mathcal{P} \bigcap \Pi_n = \phi \tag{6.3b}$$

and

$$\mathcal{P} \bigcap \Pi(\omega) = \phi \quad \forall \, \omega \epsilon \, [0, \ \infty]. \tag{6.3c}$$

Similarly, closed loop stability holds for all parameters contained in the simply connected region $\mathcal{A}$, if and only if

$$\mathcal{A} \bigcap \Pi_0 = \phi \tag{6.4a}$$

$$\mathcal{A} \bigcap \Pi_n = \phi \tag{6.4b}$$

and

$$\mathcal{A} \bigcap \Pi(\omega) = \phi \quad \forall \, \omega \epsilon [0, \ \infty]. \tag{6.4c}$$

This theorem is illustrated in Figure 6.1.

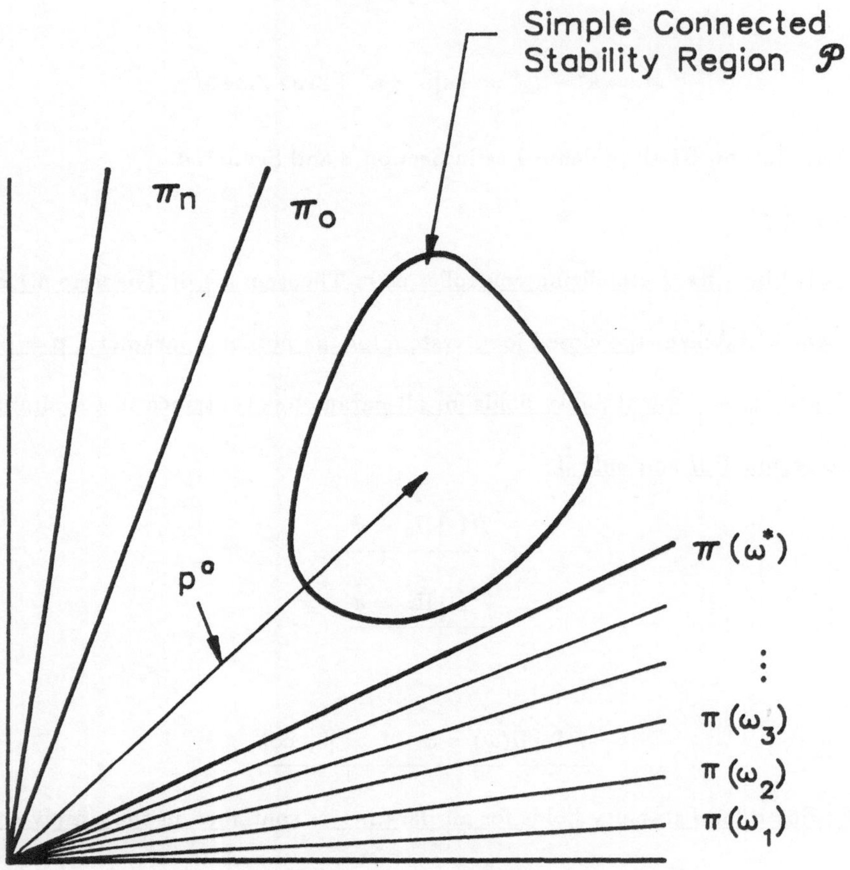

Illustration of Theorem 2.6.1

Figure 6.1

The theorem given above is a complete solution to the robust stability problem in the linear or affine cases treated in this chapter. When $\mathcal{P}$ or $\mathcal{A}$ is a regular geometric object such as a polytope or hypersphere or hyperellipsoid, the verification of the conditions (6.3) or (6.4) can be computationally relatively simple. We expect this result to play a useful role in synthesis and robustification procedures as well.

CHAPTER 3

STABILITY ELLIPSOIDS AND PERTURBATION POLYTOPES

1. INTRODUCTION

The spherical stability region calculated in the last chapter implicitly assumes that the worst case perturbation in each direction is of the same magnitude. In practice, this may not be the case, for example, when the perturbations are prescribed to lie within a rectangular polytope. This motivates us to consider weighted perturbations in the next two sections and to solve the problem of determining the largest stability ellipsoid. In Section 4, we use the stability margins calculated so far to describe a numerical procedure to design a controller for robust stability. These ideas are illustrated by some examples in section 5 which concludes with a brief discussion.

2. THE STABILITY ELLIPSOID

This section extends the method presented in the previous chapter to provide the solution to a more general problem, namely, that of finding the largest stability hyperellipsoid in the parameter space of a centered at the nominal point $\mathbf{a}^0$. As will be seen later, this is useful in applications where weighted perturbations occur.

By "largest" in the above, we mean that the shape of the ellipsoid is fixed by specifying the ratios of the principal axes as

$$\alpha_1 : \alpha_2 : \cdots : \alpha_l \tag{2.1}$$

where

$$\alpha_i > 0, \quad i = 1, 2, \ldots, l \tag{2.2}$$

and such an ellipsoid is enlarged to the maximum possible extent. Obviously, the largest stability hypersphere is a particular case of the largest stability hyperellipsoid if all α_i in (2.1) are equal.

Let $E_\epsilon(\mathbf{a}^0,\alpha)$ denote an ellipsoid centered at $\mathbf{a}^0$ with principal axes parallel to the coordinate axes and of lengths $\epsilon\alpha_1, \ldots, \epsilon\alpha_l$. Consider the family of ellipsoids

$$\mathcal{E}(\mathbf{a}^0,\alpha) := \{E_\epsilon(\mathbf{a}^0,\alpha) | 0 \leq \epsilon < \infty\} . \tag{2.3}$$

Let

$$\mathrm{Q} := \begin{pmatrix} \alpha_1 & & & & \\ & \cdot & & & \\ & & \cdot & & \\ & & & \cdot & \\ & & & & \cdot & \\ & & & & & \alpha_l \end{pmatrix} \tag{2.4}$$

and define $\tilde{\mathbf{a}}$ by

$$\mathbf{a} = \mathrm{Q}\tilde{\mathbf{a}} \tag{2.5}$$

so that

$$\tilde{\mathbf{a}} = \mathrm{Q}^{-1}\mathbf{a} . \tag{2.6}$$

Clearly, the linear transformation (2.4) maps the set of all hyperspheres in the $\tilde{\mathbf{a}}$ space, centered at $\tilde{\mathbf{a}}^0$, onto $\mathcal{E}(\mathbf{a}^0,\alpha)$ and the mapping is one-to-one. Therefore, the largest stability hyperellipsoid in the $\mathbf{a}$ space can be found by determining the largest stability hypersphere in the subsidiary space of $\tilde{\mathbf{a}}$. This can be carried out by the method described in Section 4 of Chapter 2 with the matrix

$$\tilde{\mathrm{A}} = \mathrm{AQ} \tag{2.8}$$

substituted for A, $\mathbf{a}$ replaced by $\tilde{\mathbf{a}}$, and $\mathbf{a}^0$ by $\tilde{\mathbf{a}}^0$.

Let

$$\tilde{\rho} = \tilde{\rho}(\tilde{\mathbf{a}}^0) = \min\{\tilde{r}_0, \tilde{r}_n, \tilde{r}\} \tag{2.9}$$

denote the radius of the largest stability hypersphere in the $\tilde{\mathbf{a}}$ space. The above considerations lead to the following theorem.

Theorem 2.1

Let C(s) be a given stabilizing controller as in (2.3.1) and(2.3.2). Then the largest stability hyperellipsoid $E_{\epsilon^*}(\mathbf{a}^0,\alpha)$ in the class $\mathcal{E}(\mathbf{a}^0,\alpha)$ is given by

$$\epsilon^* = \tilde{\rho}\,. \tag{2.10}$$

3. POLYTOPES OF PERTURBATIONS

In some applications, the plant parameters are known to lie within given bounds

$$a_i^0 - \gamma_i < a_i < a_i^0 + \epsilon_i, \quad i = 1, 2, \ldots, l \tag{3.1}$$

or

$$-\gamma_i < \Delta a_i < \epsilon_i, \quad i = 1, 2, \ldots l \tag{3.2}$$

and closed loop stability is required for all such values of the parameter vector. Equation (3.1) determines a rectangular polytope in the **a** space. It should be noted that whenever the parameters are perturbed independently, the stability polytope, rather than the stability hypersphere (or hyperellipsoid), is of primary interest. A procedure for treating the above problem within the framework of this chapter is to find the stability hypersphere (or hyperellipsoid) and ensure that it inscribes the polytope (3.1).

Since it is desirable to center the stability hypersphere (ellipsoid) at the center of the polytope we redefine the nominal point and the tolerances as follows. Let

$$\epsilon := [\epsilon_1, \epsilon_2, \ldots, \epsilon_l]^{\mathrm{T}}, \quad \gamma := [\gamma_1, \gamma_2, \ldots, \gamma_l]^{\mathrm{T}} \tag{3.3}$$

and introduce the new nominal parameter vector

$$\bar{\mathbf{a}}^0 = \mathbf{a}^0 + \frac{1}{2}(\epsilon - \gamma) \tag{3.4}$$

and new tolerances

$$\bar{\epsilon} = \frac{1}{2}(\epsilon + \gamma)\,. \tag{3.5}$$

Then (3.1) and (3.2) are equivalent to

$$\bar{a}_i^0 - \bar{\epsilon}_i < a_i < \bar{a}_i^0 + \bar{\epsilon}_i, \quad i = 1, \ldots l \tag{3.6}$$

and

$$-\bar{\epsilon}_i < \Delta\bar{a}_i < \bar{\epsilon}_i, \quad i = 1, \ldots l\,. \tag{3.7}$$

This shows that, for the fixed polytope problem (3.1), the perturbation classes in (1.2.7) and (1.2.8) of Section 2 Chapter 1 can both be treated within the same mathematical framework. Therefore, without loss of generality, we will consider the class (ii), i.e., we assume

$$a_i^0 - w_i\epsilon < a_i < a_i^0 + w_i\epsilon \tag{3.8}$$

where ϵ is a positive constant and w_i, $i = 1, 2, \ldots l$ are given positive weights defined in the vector form as:

$$\mathbf{w} := [w_1, w_2, \ldots, w_l]^{\mathrm{T}}\,. \tag{3.9}$$

For a fixed controller C(s) that stabilizes the nominal plant with parameter $\mathbf{a}^0$ let $\tilde{\rho}(\tilde{\mathbf{a}}^0)$ denote the radius of the largest stability hypersphere in the space $\tilde{\mathbf{a}}$ calculated according to (2.9) and let $E_{\tilde{\rho}}(\mathbf{a}^0, \alpha)$ denote the largest hyperellipsoid determined according to Theorem 2.1. Let us also define the vector

$$\mathbf{w}' := \left[\frac{w_1}{\alpha_1}, \frac{w_2}{\alpha_2}, \ldots, \frac{w_l}{\alpha_l}\right]^T \tag{3.10}$$

where α_i, $i = 1, 2, \ldots, l$ are the parameters characterizing the ellipsoid $E_{\tilde{\rho}}(\mathbf{a}^0, \alpha)$ as in Section 2.

Theorem 3.1

Let C(s) be a controller that stabilizes the plant with nominal parameter $\mathbf{a}^0$ and let $E_{\tilde{\rho}}(\mathbf{a}^0, \alpha)$ be the largest stability hyperellipsoid. Then the controller C(s) stabilizes the closed loop system for all parameters lying in the polytope (3.8) if

$$\epsilon \|\mathbf{w}'\|_2 \leq \tilde{\rho}(\tilde{\mathbf{a}}^0) \tag{3.11}$$

where $\mathbf{w}'$ is given by (3.10).

The proof of this theorem is obtained by applying transformation (2.6) to the polytope (3.8) and then inscribing the corresponding polytope into the largest stability hypersphere in the space of $\tilde{\mathbf{a}}$ centered at $\tilde{\mathbf{a}}^0$. The result of Theorem 3.1 can be strengthened by taking advantage of the fact that Π_0 and Π_n, if nonempty, are (l-1)-dimensional hyperplanes in the a space. Thus, finding the conditions ensuring that the polytope does not intersect the hyperplanes is a relatively easy task.

Theorem 3.2

Let C(s) be a controller that stabilizes the plant with nominal parameter $\mathbf{a}^0$ and let $E_{\tilde{\rho}}(\mathbf{a}^0, \alpha)$ be the largest stability hyperellipsoid. Then the controller $C(s)$ stabilizes the closed loop system for all parameters lying in the polytope (3.8) if

$$\epsilon \leq \min \left\{ \frac{|X_l A \mathbf{a}^0 + X_l \mathbf{b}|}{\sum_{i=1}^{l} |X_l A_i| w_i}, \frac{|X_f A \mathbf{a}^0 + X_f \mathbf{b}|}{\sum_{i=1}^{l} |X_f A_i| w_i}, \frac{\tilde{r}}{||\mathbf{w}'||_2} \right\} \tag{3.12}$$

where X_l and X_f, are the first and the last rows of the matrix in (2.3.8), respect–tively A_i denotes the ith column of the matrix A in (2.4.5) and $\tilde{r}$ is the same as in (2.9).

The proof of this theorem is omitted. The formula (3.12) is an improvement over (3.11) because the bound given by (3.11) can be rewritten as

$$\epsilon \leq \min \left\{ \frac{|X_l A \mathbf{a}^0 + X_l \mathbf{b}|}{||X_l A Q||_2 ||\mathbf{w}'||_2}, \frac{|X_f A \mathbf{a}^0 + X_f \mathbf{b}|}{||X_f A Q||_2 ||\mathbf{w}'||_2}, \frac{\tilde{r}}{||\mathbf{w}'||_2} \right\} \tag{3.13}$$

and it can be shown, using the Schwartz inequality, that the first two terms of (3.12) are greater than or equal to the corresponding terms in (3.13).

4. CONTROLLER DESIGN

The stability margins ρ and $\tilde{\rho}$ calculated in the previous sections are useful for comparing the robustness against variations of the parameter **a** of two given controllers. When the controller is not specfied but is to be synthesized by choosing a vector **x** of free parameters, we regard these margins as functions of **x**, i.e. $\rho = \rho(\mathbf{x})$, $\tilde{\rho} = \tilde{\rho}(\mathbf{x})$. Robustification of a controller then consists of iteratively choosing successive values of **x** to increase these margins until they are maximized, or the stability hypersphere or ellipsoid obtained contains the given range of parameter variations that are to be tolerated. The controller order is successively increased until the above criteria are met. This problem is addressed in this section. For the sake of simplicity we will describe the algorithm in the plant transfer function coefficient space **p**. The same approach, with minor modifications of the formulas involved, is applicable to the **a** space and to the $\tilde{\mathbf{a}}$ space.

The transfer function $\mathbf{C}(s)$ of the controller given by (2.3.1) and (2.3.2) specifies completely the controller parameter vector **x** in (2.3.10) or the matrix X in (2.3.8) and, conversely, the vector **x** or structured matrix X specfies the controller transfer function completely. Therefore, we refer to **x** or X as the controller and note that the distances r_0, r_n and r are functions of **x** and

$$\rho(\mathbf{x}) = \min\{r_0(\mathbf{x}),\ r_n(\mathbf{x}),\ r(\mathbf{x})\} := J(\mathbf{x})\ . \tag{4.1}$$

We define the optimization problem

$$\max_{\mathbf{x}}\ J(\mathbf{x}) = \max_{\mathbf{x}}[\min\{r_0(\mathbf{x}),\ r_n(\mathbf{x}),\ r(\mathbf{x})\}] \tag{4.2}$$

and seek to maximize $\rho(\mathbf{x})$ over all stabilizing **x**. Since a purely numerical gradient method will be used, clearly only local maxima starting from a nominal stabilizing controller will be found. Also, the min operation in the calculation of $\rho(\mathbf{x})$ means that a straightforward gradient method may not always work. It is relatively easy to sidestep this problem numerically, however, because $r_0(\mathbf{x})$, $r_n(\mathbf{x})$ and $r(\mathbf{x})$ are continuous functions of **x**.

We recall from the results of Section 3 of Chapter 2 that equations (2.3.32), (2.3.33) and (2.3.43) display the explicit dependence of r_0, r_n and r on **x**:

$$r_0^2(\mathbf{x}) = \frac{1}{X_l X_l^T}\left[\mathbf{p}^{0T} X_l^T X_l \mathbf{p}^0\right] \tag{4.3a}$$

$$r_n^2(\mathbf{x}) = \frac{1}{X_f X_f^T}[\mathbf{p^0}^T X_f^T X_f \mathbf{p^0}] \tag{4.3b}$$

and

$$r^2(\mathbf{x}) = \min_{\omega} \mathbf{p^0}^T Q_{\mathbf{x}}(\omega)\mathbf{p^0} \tag{4.3c}$$

where

$$Q_{\mathbf{x}}(\omega) = I - P_{\mathbf{x}}(\omega)(P_{\mathbf{x}}^T(\omega)P_{\mathbf{x}}(\omega))^{-1}P_{\mathbf{x}}^T(\omega) \tag{4.3d}$$

and

$$P_{\mathbf{x}}(\omega) := \begin{pmatrix} X_I^{-1}\Phi(\omega) & -X_I^{-1}X_J \\ 0 & I \end{pmatrix} \tag{4.3e}$$

The quantities X_l, X_f, X_I, X_J are specified once the controller parameter $\mathbf{x}$ is chosen.

A gradient based algorithm can be designed to update the controller by choosing the design vector $\mathbf{x}_{k+1}$ at the kth iteration so that

$$\rho(\mathbf{x}_{k+1}) \geq \rho(\mathbf{x}_k). \tag{4.4}$$

The update can be chosen as

$$\mathbf{x}_{k+1} = \mathbf{x}_k + \underbrace{\lambda \frac{\partial \rho(\mathbf{x}_k)}{\partial \mathbf{x}_k}}_{\Delta \mathbf{x}_k} \tag{4.5}$$

where the quantity $\frac{\partial \rho(\mathbf{x}_k)}{\partial \mathbf{x}_k}$ is determined by numerical means. The step size λ can be chosen to maximize $\rho(\mathbf{x}_{k+1})$ but since the closed loop system with the nominal plant and the controller $\mathbf{x}_{k+1}$ must be stable it is required that

$$P^0 \mathbf{x}_{k+1} = \delta_{k+1} \in H_n \tag{4.6}$$

where P^0 is the matrix in (2.3.9) with nominal values for the plant parameters d_i^0, $\mathbf{n}_i^0$, $i = 0, 1, ..., q$ and δ_{k+1} denotes the $(k+1)^{st}$ iterate of the vector δ. Write

$$P^0(\mathbf{x}_k + \Delta\mathbf{x}_k) = P^0\mathbf{x}_k + P^0\Delta\mathbf{x}_k = \delta_k + \Delta\delta_k\ . \tag{4.7}$$

Let $\rho(\delta_k)$ denote as in Chapter 1 the radius of the largest stability hypersphere in the characteristic vector space δ. Then the controller $\mathbf{x}_{k+1}$ preserves closed loop stability if

$$||\mathrm{P}^0 \Delta \mathbf{x}_k||_2 < \rho(\delta_k). \tag{4.8}$$

Now (4.8) can be satisfied by choosing $\Delta \mathbf{x}_k$ so that

$$||\Delta \mathbf{x}_k||_2 < \frac{\rho(\delta_k)}{||\mathrm{P}^0||_F} \tag{4.9}$$

where $||.||_F$ is the Frobenius norm. Therefore the step size λ in (4.5) can be chosen so that

$$\lambda < \frac{\rho(\delta_k)}{||\mathrm{P}^0||_F||\frac{\partial \rho(\mathbf{x}_k)}{\partial \mathbf{x}_k}||_2} \tag{4.10}$$

and this guarantees closed loop stability.

For independent perturbations of plant parameters one should choose $\mathbf{x}$ such that the allowable "box-type" perturbations are maximized (variable polytope problem). In the framework developed here this problem is best solved by choosing $\mathbf{x}$ such that the hyperellipsoid with parameters $\alpha_i = w_i$ is maximized (the shape of the ellipsoid is best fitted to the shape of the polytope).

The above robustification algorithm has been implemented and applied to various examples. Three such examples are given in the next section.

5. EXAMPLES

Example 1

As an illustrative example a satellite control problem [52, page 21] is considered. A sketch of the satellite is shown in Fig. 5.1.

For the satellite model we assume two masses connected by a spring with torque constant k and viscous damping constant d. The equations of motion from Fig. 5.1 are

$$J_1\ddot{\theta}_1 + d(\dot{\theta}_1 - \dot{\theta}_2) + k(\theta_1 - \theta_2) = T_c$$

$$J_2\ddot{\theta}_2 + d(\dot{\theta}_2 - \dot{\theta}_1) + k(\theta_2 - \theta_1) = 0$$

where T_c is the control torque and J_1 and J_2 are inertias. If we choose, as the state vector,

$$\mathbf{x}_s^{\mathrm{T}} = (\,\theta_2 \quad \dot{\theta}_2 \quad \theta_1 \quad \dot{\theta}_1\,)$$

the state equations with T_c=u and θ_1=y are

$$\dot{\mathbf{x}}_s = \begin{pmatrix} 0 & 1 & 0 & 0 \\ -\frac{k}{J_2} & -\frac{d}{J_2} & \frac{k}{J_2} & \frac{d}{J_2} \\ 0 & 0 & 0 & 1 \\ \frac{k}{J_1} & \frac{d}{J_1} & -\frac{k}{J_1} & -\frac{d}{J_1} \end{pmatrix} \mathbf{x}_s + \begin{pmatrix} 0 \\ 0 \\ 0 \\ \frac{1}{J_1} \end{pmatrix} u$$

$$y = (\,0 \quad 0 \quad 1 \quad 0\,)\,\mathbf{x}_s\,.$$

Thus, the corresponding transfer function with J_1=1 and J_2=1 is

$$G(s) = \frac{s^2 + ds + k}{s^2(s^2 + 2ds + 2k)}\,. \tag{5.1}$$

Physical analysis of the boom leads to the conclusion that the k and d parameters vary within bounds given by

$$0.09 \le k \le 0.4 \tag{5.2}$$

$$0.04\sqrt{\frac{k}{10}} \le d \le 0.2\sqrt{\frac{k}{10}}\,. \tag{5.3}$$

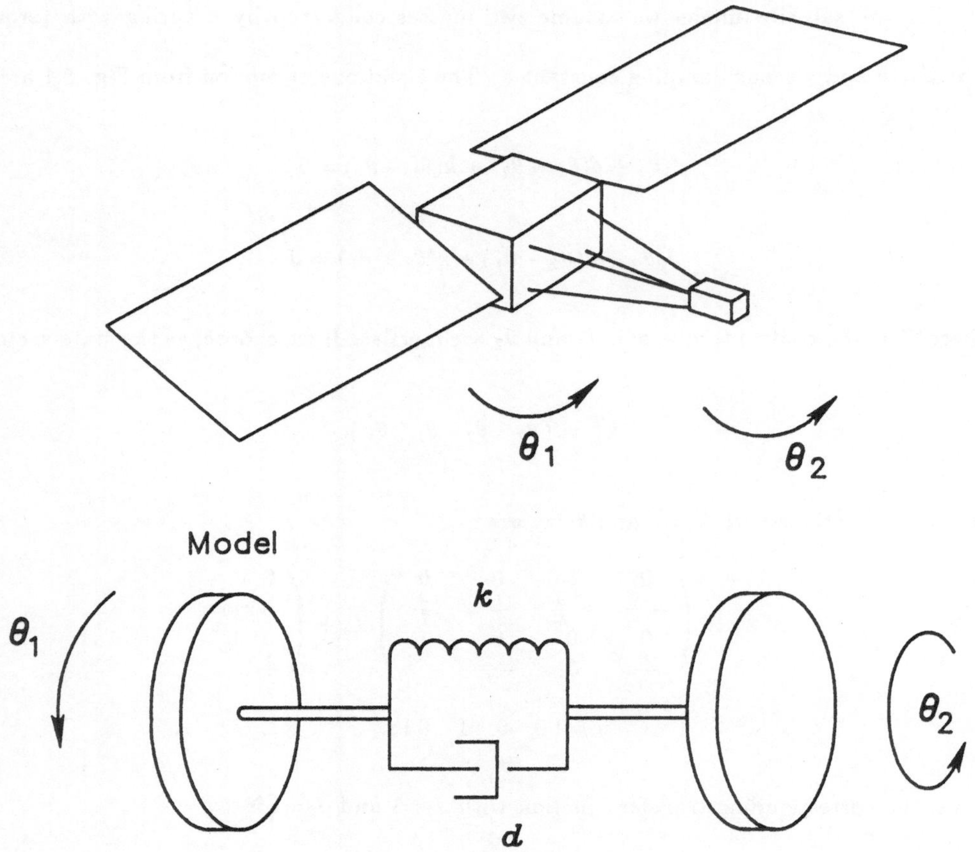

Figure 5.1

A sketch of the satellite

As a result, the vehicle resonance ω_n can vary between 1 and 2 rad/s, and the damping ratio ζ varies between 0.02 and 0.1. Our problem is to design a controller which stabilizes the closed loop system for all perturbations given by (5.2) and (5.3).

We select k^0=0.245 and d^0=0.0218973 which are the middle points of the variation ranges of (5.2) and (5.3) as nominal values. Therefore, the nominal vector $\mathbf{a}^0$ of physical plant parameters which are subject to perturbation is

$$\mathbf{a}^0 = \begin{pmatrix} k^0 \\ d^0 \end{pmatrix}.$$

Our objective is to obtain a low order controller $\mathbf{x}$ which will generate a stability hyper-ellipsoid inscribing the perturbation bound given by (5.2) and (5.3). We start with the stability hypersphere of radius $\rho(\mathbf{a}^0)$ in the space of physical plant parameters $\mathbf{a}$ which is centered at $\mathbf{a}^0$ and with a 0th order controller. The controller is

$$\mathrm{C}(s) = \frac{n_{c0}}{d_{c0}}$$

and the closed loop characteristic vector is

$$\underbrace{\begin{pmatrix} 0 & d_4 \\ 0 & d_3 \\ n_2 & d_2 \\ n_1 & 0 \\ n_0 & 0 \end{pmatrix}}_{\mathbf{P}} \underbrace{\begin{pmatrix} n_{c0} \\ d_{c0} \end{pmatrix}}_{\mathbf{x}} = \delta$$

or

$$\underbrace{\begin{pmatrix} 0 & 0 & 0 & 0 & d_{c0} & 0 \\ 0 & 0 & 0 & d_{c0} & 0 & 0 \\ 0 & 0 & d_{c0} & 0 & 0 & n_{c0} \\ 0 & n_{c0} & 0 & 0 & 0 & 0 \\ n_{c0} & 0 & 0 & 0 & 0 & 0 \end{pmatrix}}_{\mathbf{X}} \underbrace{\begin{pmatrix} n_0 \\ n_1 \\ d_2 \\ d_3 \\ d_4 \\ n_2 \end{pmatrix}}_{\mathbf{p}} = \delta.$$

$$X_I = \begin{pmatrix} 0 & 0 & 0 & 0 & d_{c0} \\ 0 & 0 & 0 & d_{c0} & 0 \\ 0 & 0 & d_{c0} & 0 & 0 \\ 0 & n_{c0} & 0 & 0 & \\ n_{c0} & 0 & 0 & 0 & 0 \end{pmatrix}, \quad X_J = \begin{pmatrix} 0 \\ 0 \\ n_{c0} \\ 0 \\ 0 \end{pmatrix}$$

From (5.1) there is the following linear relation between the transfer function coefficients of plant n_i and d_i, and physical parameters, d and k.

$$\underbrace{\begin{pmatrix} n_0 \\ n_1 \\ d_2 \\ d_3 \\ d_4 \\ n_2 \end{pmatrix}}_{\mathbf{p}} = \underbrace{\begin{pmatrix} 1 & 0 \\ 0 & 1 \\ 2 & 0 \\ 0 & 2 \\ 0 & 0 \\ 0 & 0 \end{pmatrix}}_{\mathbf{A}} \underbrace{\begin{pmatrix} k \\ d \end{pmatrix}}_{\mathbf{a}} + \underbrace{\begin{pmatrix} 0 \\ 0 \\ 0 \\ 0 \\ 1 \\ 1 \end{pmatrix}}_{\mathbf{b}} .$$

From (2.4.12) and (2.4.14) we obtain

$$r_0 = k^0 \ , \ r_n = \infty$$

since $X_l A = [n_{c0}, 0]$, $X_l b = 0$ and $X_f A = [0, 0]$.

Now, following (2.4.15)-(2.4.20), we get the solution (2.4.17)

$$\mathbf{t_a}(\omega) = \begin{pmatrix} \mathbf{t}_k(\omega) \\ \mathbf{t}_d(\omega) \end{pmatrix} = \begin{pmatrix} l_0\omega^2/n_{c0} \\ l_1\omega^2/n_{c0} \end{pmatrix}$$

and then we formulate equation (2.4.20) whose solution is

$$l_2 = d_{c0}, \ l_1 = 0, \ l_0 = n_{c0}\frac{d_{c0}\omega^2 - n_{c0}}{2d_{c0}\omega^2 - n_{c0}}, \ \mathbf{t}_J = 1$$

if $\omega \neq \sqrt{n_{c0}/2d_{c0}}$. For $\omega = \sqrt{n_{c0}/2d_{c0}}$ equation (2.4.20) is inconsistent and therefore

$$\Omega = [(0, \infty) \backslash \{\sqrt{n_{c0}/2d_{c0}}\}] .$$

For $\omega \in \Omega$, $\mathbf{t}(\omega) \in \Pi(\omega)$ takes the form (equation (2.4.24))

$$\begin{pmatrix} \mathbf{t}_k(\omega) \\ \mathbf{t}_d(\omega) \end{pmatrix} = \begin{pmatrix} \omega^2 \frac{d_{c0}\omega^2 - n_{c0}}{2d_{c0}\omega^2 - n_{c0}} \\ 0 \end{pmatrix} .$$

Note that $\tilde{P}(\omega)$ and $\bar{I}_t$ of (2.4.24) do not appear here so finally

$$r^2(\omega) = \begin{cases} \infty & \text{if } \omega = \sqrt{n_{c0}/2d_{c0}}, \\ \left(k^0 - \omega^2 \frac{d_{c0}\omega^2 - n_{c0}}{2d_{c0}\omega^2 - n_{c0}}\right)^2 & +(d^0)^2 \text{ otherwise .} \end{cases}$$

It can be easily shown that $r^2 = \min_{0 \leq \omega < \infty} r^2(\omega) = (d^0)^2$, so $r = d^0$ and, since $d^0 < k^0$,

$$\rho(\mathbf{a}^0) = d^0.$$

It is interesting that the above result does not depend on the controller $\mathbf{x}$, i.e., it holds for any stabilizing controller C(s) of order zero.

From Fig. 5.2 we see that the stability hypersphere S_0, with radius $\rho(\mathbf{a}^0)$, does not inscribe the given perturbation bound. Because of the oblong range of the perturbation region, we now consider an ellipsoid $E_\epsilon(\mathbf{a}^0, \alpha)$ in the space of k and d, centered at (k^0, d^0), with $\alpha = [1, \alpha_d]$. The polytope which is to be inscribed into the ellipse is given by

$$-\epsilon < \Delta k < \epsilon$$

$$-0.1168\epsilon < \Delta d < 0.1168\epsilon$$

where $\epsilon = 0.155$ and $\mathbf{w} = [1, 0.1168]$. Now, from Theorem 3.2 we have the condition

$$\epsilon \leq \min \left\{ k^0, \infty, \frac{\tilde{r}}{\sqrt{1 + (\frac{0.1168}{\alpha_d})^2}} \right\}$$

where it can be shown that $\tilde{r} = \tilde{d}^0 = d^0/\alpha_d$. The above condition can be satisfied by any $\alpha_d < 0.079$. For example, for $\alpha_d = 0.07$ we obtain the ellipse E with the semiaxes 0.3128186 and 0.0218973 containing the polytope given, as shown in Fig. 5.2.

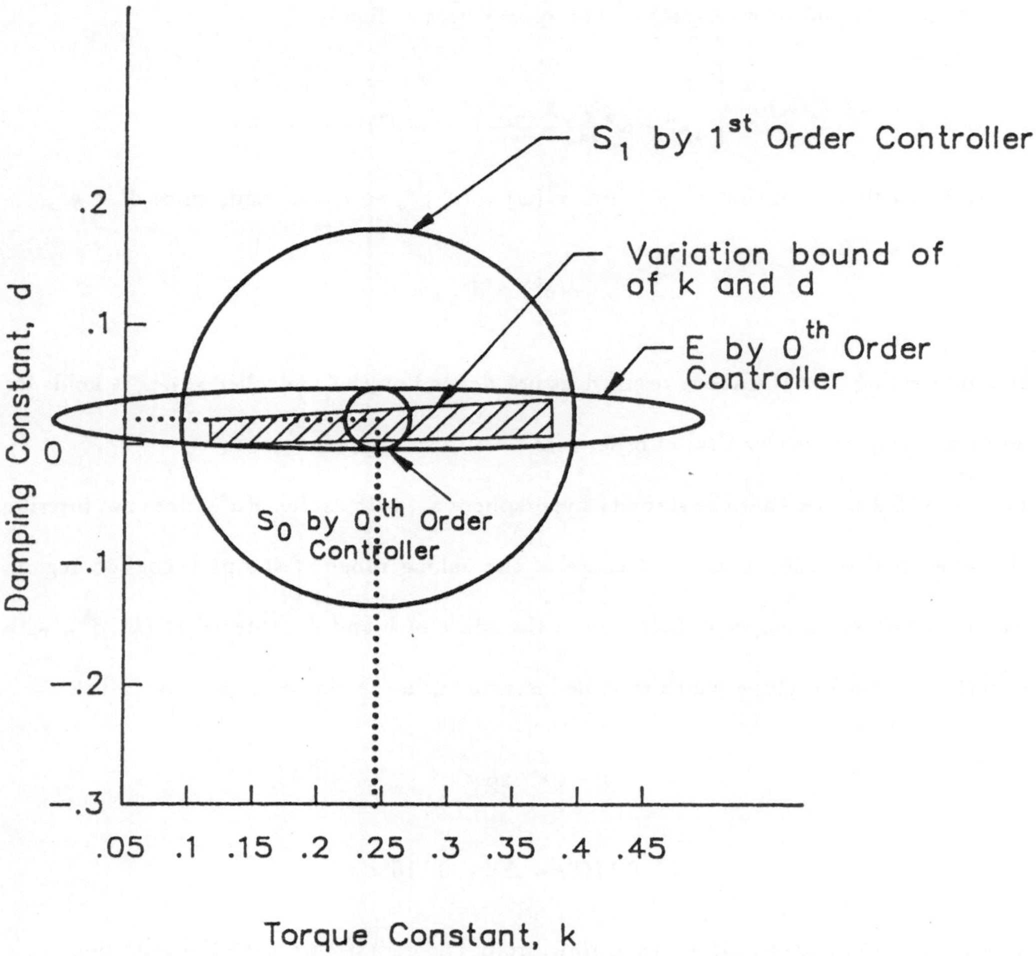

Figure 5.2

Perturbation bounds and stability regions in k and d space.

Note that the above solution is independent of the controller as long as $C(s)$ is of 0th order and stabilizes the plant. Therefore every 0th order stabilizing controller will be robust for the perturbations given and the value n_{c0}/d_{c0} can be used to satisfy other design requirements. In fact, it can be shown for the plant considered that for any 0th order stabilizing controller, i.e., such that $n_{c0}/d_{c0} > 0$, the parameters k and d can be perturbed anywhere in the first quadrant. We have also calculated the largest stability hypersphere obtained for a "maximally" robust first order controller using the algorithm of Section 4 and this is shown in Fig. 5.2. as S_1.

Example 2

As another illustrative example a control problem of a digital tape transport system [52, page 504] is considered. The model of a tape drive is shown in Fig. 5.3.

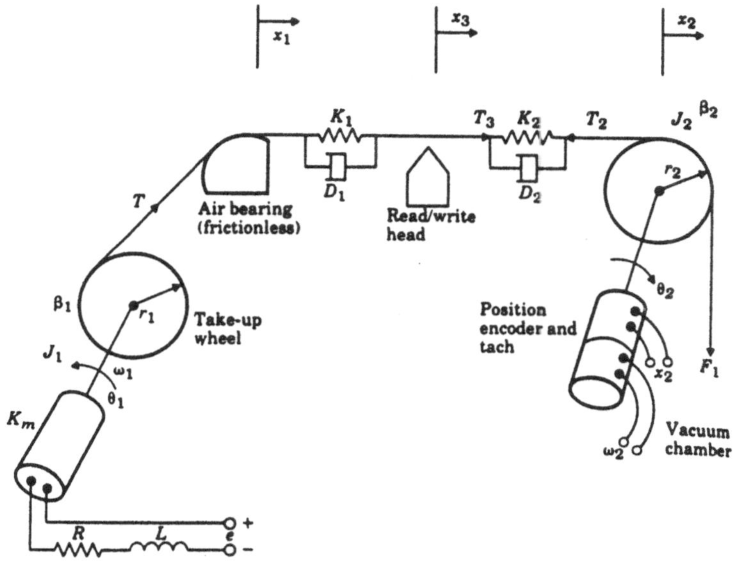

Figure 5.3

The model of a tape drive

The system is in static equilibrium when $T_0 = F_1$ and $K_m i_0 = r_1 T_0$. We define the variables as deviations from this equilibrium. The equations of motion of the system are given by the laws of mechanics:

$$J_1 \frac{d\omega_1}{dt} + \beta_1 \omega_1 - r_1 T = K_m i$$

$$\dot{x}_1 = r_1 \omega_1$$

$$L \frac{di}{dt} + Ri + K_m \omega_1 = e$$

$$\dot{x}_2 = r_2 \omega_2$$

$$J_2 \frac{d\omega_2}{dt} + \beta_2 \omega_2 = -r_2 T$$

$$T = K_1(x_3 - x_1) + D_1(\dot{x}_3 - \dot{x}_1)$$

$$T = K_2(x_2 - x_3) + D_2(\dot{x}_2 - \dot{x}_3)$$

$$x_1 = r_1\theta_1 \quad , x_2 = r_2\theta_2 \quad , x_3 = \frac{x_1 + x_2}{2}.$$

Assuming that $K_1{=}K_2{:=}K$, $D_1{=}D_2{:=}D$ and substituting the numerical values given in [52], we obtain the following transfer function

$$\mathrm{G}(s) = \frac{X_2(s)}{E_1(s)} = \frac{n_1 s + n_0}{d_5 s^5 + d_4 s^4 + d_3 s^3 + d_2 s^2 + d_1 s + d_0}$$

where,

$n_1 = 0.03D$

$n_0 = 0.3 \times 10^{-4} K$

$d_5 = 1$

$d_4 = 0.025D + 2.25$

$$d_3 = 0.035D + 0.25 \times 10^{-4}K + 1.5225$$

$$d_2 = 0.01045D + 0.35 \times 10^{-4}K + 0.2725$$

$$d_1 = 0.1045 \times 10^{-4}K$$

$$d_0 = 0 \ .$$

We assume that K and D are subject to perturbation.With nominal values D^0=20 and $K^0 = 4 \times 10^4$

$$G(s) = \frac{0.6s + 1.2}{s^5 + 2.75s^4 + 3.2225s^3 + 1.8815s^2 + 0.418s} \ .$$

Our aim is to find stabilizing controllers which maximize the radius of the stability hypersphere in the space of plant primary parameters K and D. We start with a 0th order controller

$$C(s) = \frac{n_{c0}}{d_{c0}}.$$

The closed loop characteristic vector is

$$\underbrace{\begin{pmatrix} 0 & d_5 \\ 0 & d_4 \\ 0 & d_3 \\ 0 & d_2 \\ n_1 & d_1 \\ n_0 & 0 \end{pmatrix}}_{\mathbf{P}} \underbrace{\begin{pmatrix} n_{c0} \\ d_{c0} \end{pmatrix}}_{\mathbf{x}} = \delta$$

or

$$\underbrace{\begin{pmatrix} 0 & 0 & 0 & 0 & 0 & d_{c0} & 0 \\ 0 & 0 & 0 & 0 & d_{c0} & 0 & 0 \\ 0 & 0 & 0 & d_{c0} & 0 & 0 & 0 \\ 0 & 0 & d_{c0} & 0 & 0 & 0 & 0 \\ 0 & n_{c0} & 0 & 0 & 0 & 0 & d_{c0} \\ n_{c0} & 0 & 0 & 0 & 0 & 0 & 0 \end{pmatrix}}_{\mathbf{X}} \underbrace{\begin{pmatrix} n_0 \\ n_1 \\ d_2 \\ d_3 \\ d_4 \\ d_5 \\ d_1 \end{pmatrix}}_{\mathbf{P}} = \delta$$

$$\mathrm{X}_I = \begin{pmatrix} 0 & 0 & 0 & 0 & 0 & d_{c0} \\ 0 & 0 & 0 & 0 & d_{c0} & 0 \\ 0 & 0 & 0 & d_{c0} & 0 & 0 \\ 0 & 0 & d_{c0} & 0 & 0 & 0 \\ 0 & n_{c0} & 0 & 0 & 0 & 0 \\ n_{c0} & 0 & 0 & 0 & 0 & 0 \end{pmatrix}, \quad \mathrm{X}_J = \begin{pmatrix} 0 \\ 0 \\ 0 \\ 0 \\ d_{c0} \\ 0 \end{pmatrix}.$$

We have

$$\underbrace{\begin{pmatrix} n_0 \\ n_1 \\ d_2 \\ d_3 \\ d_4 \\ d_5 \\ d_1 \end{pmatrix}}_{\mathbf{p}} = \underbrace{\begin{pmatrix} 0 & 0.3 \times 10^{-4} \\ 0.03 & 0 \\ 0.01045 & 0.35 \times 10^{-4} \\ 0.035 & 0.25 \times 10^{-4} \\ 0.025 & 0 \\ 0 & 0 \\ 0 & 0.1045 \times 10^{-4} \end{pmatrix}}_{\mathbf{A}} \underbrace{\begin{pmatrix} D \\ K \end{pmatrix}}_{\mathbf{a}} + \underbrace{\begin{pmatrix} 0 \\ 0 \\ 0.2725 \\ 1.5225 \\ 2.25 \\ 1 \\ 0 \end{pmatrix}}_{\mathbf{b}}.$$

Now, following (2.4.15)–(2.4.24) we obtain

$$\mathrm{t}(\omega) = \begin{pmatrix} \mathrm{t}_D(\omega) \\ \mathrm{t}_K(\omega) \end{pmatrix} = \frac{1}{ad - bc} \begin{pmatrix} de - bf \\ af - ce \end{pmatrix}$$

where

$$a = 0.03 \tfrac{n_{c0}}{\omega^2} - 0.035 d_{c0}$$

$$b = 0.1045 \times 10^{-4} \tfrac{d_{c0}}{\omega^2} - 0.25 \times 10^{-4} d_{c0}$$

$$c = -0.01045 d_{c0} + 0.025 d_{c0} \omega^2$$

$$d = 0.3 \times 10^{-4} \tfrac{n_{c0}}{\omega^2} - 0.35 \times 10^{-4} d_{c0}$$

$$e = 1.5225 d_{c0} - \omega^2 d_{c0}$$

$$f = 0.2725 d_{c0} - 2.25 \omega^2 d_{c0}$$

Now, we get

$$r^2(\omega) = (D^0 - \mathrm{t}_D(\omega))^2 + (K^0 - \mathrm{t}_K(\omega))^2 .$$

Applying the optimization procedure of Section 4 we obtained the following:

$$\mathrm{C}(s) = 0.1/4.2214 = 0.02368$$

and

$$r^2 = \min_{\omega} r^2(\omega) = 2875.9946$$

Therefore

$$\rho = \min\{K^0, \infty, r\} = 53.618$$

is the stability margin obtained. Next, we wish to find a low order controller which will generate the largest stability ellipsoid. For this ellipsoid two weighting constants are chosen as follows based on the relative magnitudes of D^0 and K^0:

$$\alpha_D = 1, \ \alpha_K = 2000$$

Similarly, we get

$$\bar{r}^2(\omega) = \left(\frac{D^0}{\alpha_D} - \frac{t_D(\omega)}{\alpha_D}\right)^2 + \left(\frac{K^0}{\alpha_K} - \frac{t_K(\omega)}{\alpha_K}\right)^2$$

We obtained the following robust controller:

$$C(s) = 0.1/12.078 = 0.00827$$

and

$$\bar{r}^2 = \min_{\omega} \bar{r}^2(\omega) = 99.801$$

Therefore,

$$\bar{\rho} = \min\left\{\frac{4 \times 10^4}{\alpha_k}, \bar{r}, \infty\right\} = 9.99$$

is the stability margin. This gives about 35% allowable relative perturbations.

Now we increase the order of the controller from 0 to 1. The controller transfer function is

$$C(s) = \frac{n_{c1}s + n_{c0}}{d_{c1}s + d_{c0}} \ .$$

Now P, X, p, x and X_I are given by

$$\mathbf{P} = \begin{pmatrix} 0 & d_5 & 0 & 0 \\ 0 & d_4 & 0 & d_5 \\ 0 & d_3 & 0 & d_4 \\ 0 & d_2 & 0 & d_3 \\ n_1 & d_1 & 0 & d_2 \\ n_0 & 0 & n_1 & d_1 \\ 0 & 0 & n_0 & 0 \end{pmatrix}, \quad \mathbf{x} = \begin{pmatrix} n_{c1} \\ d_{c1} \\ n_{c0} \\ d_{c0} \end{pmatrix}$$

$$\mathbf{X} = \mathbf{X}_I = \begin{pmatrix} 0 & 0 & 0 & 0 & 0 & d_{c1} & 0 \\ 0 & 0 & 0 & 0 & d_{c1} & d_{c0} & 0 \\ 0 & 0 & 0 & d_{c1} & d_{c0} & 0 & 0 \\ 0 & 0 & d_{c1} & d_{c0} & 0 & 0 & 0 \\ 0 & n_{c1} & d_{c0} & 0 & 0 & 0 & d_{c1} \\ n_{c1} & n_{c0} & 0 & 0 & 0 & 0 & d_{c0} \\ n_{c0} & 0 & 0 & 0 & 0 & 0 & 0 \end{pmatrix}, \quad \mathbf{p} = \begin{pmatrix} n_0 \\ n_1 \\ d_2 \\ d_3 \\ d_4 \\ d_5 \\ d_1 \end{pmatrix}$$

and the relation $\mathbf{p} = \mathbf{A}\mathbf{a} + \mathbf{b}$ is the same as before. We get

$$\mathbf{t_a}(\omega) = \begin{pmatrix} \mathbf{t}_D(\omega) \\ \mathbf{t}_K(\omega) \end{pmatrix} = \frac{1}{ad - bc} \begin{pmatrix} de - bf \\ af - ce \end{pmatrix}$$

where

$$a = 0.025 d_{c1}\omega^2 + 0.03 \frac{n_{c0}}{\omega^2} - 0.035 d_{c0} - 0.01045 d_{c1}$$

$$b = 0.3 \times 10^{-4} \frac{n_{c1}}{\omega^2} + 0.1045 \times 10^{-4} \frac{d_{c0}}{\omega^2} - 0.25 \times 10^{-4} d_{c0} - 0.35 \times 10^{-4} d_{c1}$$

$$c = 0.025 d_{c0}\omega^2 + 0.035 d_{c1}\omega^2 - 0.03 n_{c1} - 0.01045 d_{c0}$$

$$d = 0.3 \times 10^{-4} \frac{n_{c0}}{\omega^2} + 0.25 \times 10^{-4} \omega^2 d_{c1} - 0.35 \times 10^{-4} d_{c0} - 0.1045 \times 10^{-4} d_{c1}$$

$$e = 1.5225 d_{c0} + 0.2725 d_{c1} - \omega^2 d_{c0} - 2.25\omega^2 d_{c1}$$

$$f = 0.2725 d_{c0} - 2.25\omega^2 d_{c0} - 1.5225\omega^2 d_{c1} + \omega^4 d_{c1}$$

and

$$r^2(\omega) = (D^0 - \mathbf{t}_D(\omega))^2 + (K^0 - \mathbf{t}_K(\omega))^2 .$$

Again, using the algorithm of Section 4, we obtained the following robust controller

$$C(s) = \frac{0.1s + 0.12}{2.65s + 0.901}$$

and

$$r^2 = \min_{\omega} r^2(\omega) = 3061.7409$$

Therefore

$$\rho = \min\left\{k^0, \infty, r\right\} = 55.333$$

is the stability margin. With the same weighting coefficients as in the 0th order controller, we obtained the robust controller which generates the largest stability ellipsoid

$$C(s) = \frac{0.1s + 0.1}{4.204s + 0.9296}$$

and

$$\bar{r}^2 = \min_{0 \leq \omega < \infty} \bar{r}^2(\omega) = 158.0484 \ .$$

Therefore

$$\bar{\rho} = \min\left\{\frac{4 \times 10^4}{\alpha_K}, \infty, \bar{r}\right\} = 12.57173$$

is the stability margin which gives more than about 44% allowable relative perturbations.

Example 3

As another illustrative example a control problem of a VTOL helicopter,[53],is considered. The linearized model of the vehicle in the vertical plane is described by the state equations

$$\dot{\mathbf{x}} = A\mathbf{x} + B\mathbf{u} \ , \ y = C\mathbf{x} \ .$$

The state vector $\mathbf{x} \in R^4$ and the components of $\mathbf{x}$ are:

x_1 : horizontal velocity (knots),

x_2 : vertical velocity (knots),

x_3 : pitch rate (degrees/s),

x_4 : pitch angle (degrees).

The control vector $\mathbf{u}=[u_1, u_2]^T$, where

u_1 : collective pitch control,

u_2 : longitudinal cyclic pitch control.

Following [53], the matrices A, B and C for typical loading and flight conditions at the airspeed of 135 knots are:

$$A = \begin{pmatrix} -0.0366 & 0.0271 & 0.0188 & -0.4555 \\ 0.0482 & -1.01 & 0.0024 & -4.0208 \\ 0.1002 & a_{32} & -0.707 & a_{34} \\ 0 & 0 & 1 & 0 \end{pmatrix}$$

$$B = \begin{pmatrix} 0.4422 & 0.1761 \\ b_{21} & -7.5922 \\ -5.52 & 4.49 \\ 0 & 0 \end{pmatrix}$$

$$C = (0\ 1\ 0\ 0)\ .$$

As the airspeed changes, all the elements of the first three rows of both matrices also change. The most significant changes take place in the elements a_{32}, a_{34} and b_{21}. Therefore, in the following, all the other elements are assumed to be constants. The nominal values of a_{32}, a_{34} and b_{21} are 0.3681, 1.42 and 3.5446, respectively. The bounds on the variation of the parameters depend on the desired equilibrium state. The following bounds on the parameters are given in [53] for linear controls:

$$|\Delta a_{32}| \leq 0.05, \ |\Delta a_{34}| \leq 0.01, \ |\Delta b_{21}| \leq 0.04\ . \tag{5.4}$$

Given these bounds on the parameters , we use our approach to find a controller which stabilizes the family of plants given by (5.4). The transfer function of the system is

$$\mathbf{G}(s) = \frac{1}{d(s)}[n_1(s), n_2(s)]$$

with

$$d(s) = d_4 s^4 + d_3 s^3 + d_2 s^2 + d_1 s + d_0$$

$$n_1(s) = n_{31} s^3 + n_{21} s^2 + n_{11} s + n_{01}$$

$$n_2(s) = n_{32} s^3 + n_{22} s^2 + n_{12} s + n_{02}$$

where

$d_4 = 1$

$d_3 = 1.7536$

$d_2 = -a_{34} - 0.0024\ a_{32} + 0.7737$

$d_1 = -1.0466\ a_{34} + 4.0198\ a_{32} + 0.0689$

$d_0 = -0.0356\ a_{34} + 0.1691\ a_{32} + 0.0570$

$n_{31} = b_{21}$

$n_{21} = 0.7436\ b_{21} + 0.0080$

$n_{11} = -b_{21}\ a_{34} + 0.0239\ b_{21} + 22.2045$

$n_{01} = -0.0366\ b_{21}\ a_{34} - 0.0213\ a_{34} + 0.0456\ b_{21} + 0.7553$

$n_{32} = -7.5922$

$n_{22} = -5.6262$

$n_{12} = 7.5922\ a_{34} - 18.2250$

$n_{02} = 0.2693\ a_{34} - 1.1767$.

Now we select the vector **a** in (4.1) to "linearize" the problem:

$$\mathbf{a} = \begin{pmatrix} a_{32} \\ a_{34} \\ b_{21} \\ b_{21} a_{34} \end{pmatrix} .$$

Then A and **b** of (2.4.5) are given by

$$A = \begin{pmatrix} 0 & -0.213 & 0.0456 & -0.0366 \\ 0 & 0.2693 & 0 & 0 \\ 0.1691 & -0.0356 & 0 & 0 \\ 0 & 0 & 0.0239 & -1 \\ 0 & 7.5922 & 0 & 0 \\ 4.0198 & -1.0466 & 0 & 0 \\ 0 & 0 & 0.7436 & 0 \\ 0 & 0 & 0 & 0 \\ -0.0024 & -1 & 0 & 0 \\ 0 & 0 & 1 & 0 \\ 0 & 0 & 0 & 0 \\ 0 & 0 & 0 & 0 \\ 0 & 0 & 0 & 0 \end{pmatrix}, \quad \mathbf{b} = \begin{pmatrix} 0.7553 \\ -1.1767 \\ 0.0570 \\ 22.2045 \\ -18.2250 \\ 0.0689 \\ 0.0080 \\ -5.626 \\ 0.7737 \\ 0 \\ -7.5922 \\ 1.7536 \\ 1 \end{pmatrix}.$$

The nominal value of **a** is

$$\mathbf{a}^0 = \begin{pmatrix} 0.3681 \\ 1.42 \\ 3.5446 \\ 5.0333 \end{pmatrix}$$

and the perturbation bounds of **a** are derived from (5.4) as

$$\begin{aligned} a_1^0 - 0.05 \le a_1 \le a_1^0 + 0.05 \\ a_2^0 - 0.01 \le a_2 \le a_2^0 + 0.01 \\ a_3^0 - 0.04 \le a_3 \le a_3^0 + 0.04 \\ a_4^0 - 0.0918 \le a_4 \le a_4^0 + 0.0926 \, . \end{aligned} \qquad (5.5)$$

Using the procedure of Section 4 we get the following 0th order controller and stability margin

$$\mathbf{C}(s) = \frac{1}{0.104} \begin{pmatrix} 0.213 \\ -0.331 \end{pmatrix}$$

$$\rho_0 = 1.276 \, .$$

Assuming $w'_i = 1, \ i = 1,..,4$ in Theorem 3.1 we get $\epsilon \leq \frac{1.276}{2} = 0.638$ which shows that the stability hypersphere contains the polytope of primary parameter perturbations (5.5). Therefore closed loop stability is guaranteed with this controller for all perturbations of the original parameters given by (5.4).

Discussion of the examples

In example 1, the largest stability hypersphere for a 0th order controller is found to be independent of the controller parameters. Since this hypersphere did not contain the perturbation region, we attempted to solve the problem with a 1st order controller. This was successful because the corresponding stability hypersphere did contain the perturbation region (Fig.5.2). However, it was interesting to observe, that by properly shaping an ellipsoid adjusted to the perturbation region, robust stabilization could be achieved with a 0th order controller also. Example 2 shows that tolerance of larger perturbation ranges can be obtained by increasing the controller order, because the corresponding stability hyperspheres and ellipsoids enlarge significantly. Example 3 shows how nonlinear combinations of parameters can be handled, by introducing primary parameters to "linearize" the problem, transforming the original perturbation polytope to a corresponding one in the space of the primary parameters and inscribing the latter into a stability hypersphere or hyperellipsoid constructed, in this space, by choice of a controller.

CHAPTER 4

ROBUST STABILIZATION:THE GENERAL CASE

1. INTRODUCTION

In this chapter, we continue our treatment of the structured robust stability and robust stabilization problem via analysis of the closed loop characteristic polynomial. Here, we drop the assumption made in Chapters 2 and 3 that the parameter vector enters the plant transfer function coefficients linearly or affinely, and let this functional dependence be arbitrary.

In the next section, we give formulas for the characteristic polynomial in a particular "linearized" form that are applicable to MIMO systems. Based on this, a stability margin that measures the robustness of a given controller is defined in Section 3. A robustification procedure that attempts to increase this margin by redesigning the controller parameter vector is described in Section 4. The definition of the stability margin is such that if the margin exceeds a certain prescribed number, the corresponding controller guarantees closed loop stability for the class of perturbations given. The results are illustrated by an example in Section 5.

2. CHARACTERISTIC POLYNOMIAL CALCULATION

Consider the standard multivariable feedback system in Fig.2.1a with the plant transfer matrix G(s) and the feedback controller C(s). Let **p** be a vector of real physical plant parameters that is subject to uncertainty. In general the coefficients of the entries of the transfer matrix G(s) will be nonlinear functions of the parameter **p**. In this section we display a special form of the closed loop characteristic polynomial on which the transfer function design procedure will be based.

Let n be the order of the plant (Mcmillan degree of G(s)), t the order of the feedback controller C(s) and let

$$\delta(s) = \delta_{n+t}s^{n+t} + \cdots\cdots + \delta_1 s + \delta_0 \tag{2.1}$$

denote the closed loop characteristic polynomial. We refer as before to

$$\delta := (\ \delta_{n+t} \quad \delta_{n+t-1} \quad \cdots \quad \delta_1 \quad \delta_0\ )^T \tag{2.2}$$

as the characteristic vector and for convenience say that δ is Hurwitz if and only if $\delta(s)$ is Hurwitz.

In the following, we show how to define the controller parameter vector **x** so that the closed loop characteristic vector δ satisfies the equation

$$\mathbf{M}(\mathbf{p})c(\mathbf{x}) = \delta \tag{2.3}$$

where $\mathbf{M}(\mathbf{p})$ is a matrix containing only plant parameters and $c(\mathbf{x})$ is a vector containing only the controller parameter vector **x**. The vector **x** is a quantity that completely defines the tth order controller and conversely any t^{th} order controller determines **x**.

2.1 Single Input Single Output(SISO) systems

Let **p** denote a vector of parameters that enters the plant transfer function coefficients. Then

$$\mathbf{G}(s) = \frac{n_n(\mathbf{p})s^n + n_{n-1}(\mathbf{p})s^{n-1} + \cdots\cdots + n_1(\mathbf{p})s + n_0(\mathbf{p})}{d_n(\mathbf{p})s^n + d_{n-1}(\mathbf{p})s^{n-1} + \cdots\cdots + d_1(\mathbf{p})s + d_0(\mathbf{p})} \tag{2.4}$$

$$\mathbf{C}(s) = \frac{\beta_t s^t + \beta_{t-1}s^{t-1} + \cdots\cdots + \beta_1 s + \beta_0}{\alpha_t s^t + \alpha_{t-1}s^{t-1} + \cdots\cdots + \alpha_1 s + \alpha_0}\ . \tag{2.5}$$

With

$$c(\mathbf{x}) := (\,\alpha_t \quad \alpha_{t-1} \quad \cdots \quad \alpha_0 \quad \beta_t \quad \beta_{t-1} \quad \cdots \quad \beta_0\,)^T := \mathbf{x} \tag{2.6}$$

we get the closed loop characteristic vector δ given by

$$\underbrace{\begin{pmatrix} d_n(\mathbf{p}) & 0 & \cdots & \cdots & n_n(\mathbf{p}) & 0 & \cdots & \cdots \\ d_{n-1}(\mathbf{p}) & d_n(\mathbf{p}) & \ddots & \cdots & n_{n-1}(\mathbf{p}) & n_n(\mathbf{p}) & \ddots & \cdots \\ \vdots & d_{n-1}(\mathbf{p}) & \ddots & \cdots & \vdots & n_{n-1}(\mathbf{p}) & \ddots & \cdots \\ \vdots & \vdots & \ddots & \ddots & \vdots & \vdots & \ddots & \ddots \\ d_0(\mathbf{p}) & \vdots & \ddots & \ddots & n_0(\mathbf{p}) & \vdots & \ddots & \ddots \\ 0 & d_0(\mathbf{p}) & \ddots & \ddots & 0 & n_0(\mathbf{p}) & \ddots & \ddots \\ \vdots & \vdots & \ddots & \ddots & \vdots & \vdots & \ddots & \ddots \end{pmatrix}}_{\mathbf{M}(\mathbf{p})} \underbrace{\begin{pmatrix} \alpha_t \\ \alpha_{t-1} \\ \vdots \\ \alpha_0 \\ \beta_t \\ \vdots \\ \beta_0 \end{pmatrix}}_{c(\mathbf{x})}$$

$$= \underbrace{\begin{pmatrix} \delta_{n+t} \\ \delta_{n+t-1} \\ \vdots \\ \vdots \\ \vdots \\ \delta_1 \\ \delta_0 \end{pmatrix}}_{\delta} . \tag{2.7}$$

2.2 Single Input Multioutput Systems

Let

$$\mathbf{G}(s) = (\,\mathbf{G}_1(s) \quad \mathbf{G}_2(s) \quad \cdots \quad \cdots \quad \mathbf{G}_m(s)\,)^T \tag{2.8}$$

$$\mathbf{C}(s) = (\,\mathbf{C}_1(s) \quad \mathbf{C}_2(s) \quad \cdots \quad \cdots \quad \mathbf{C}_m(s)\,) \tag{2.9}$$

where

$$\mathbf{G}_i(s) = \frac{n_{in}(\mathbf{p})s^n + n_{in-1}(\mathbf{p})s^{n-1} + \cdots\cdots + n_{i1}(\mathbf{p})s + n_{i0}(\mathbf{p})}{d_n(\mathbf{p})s^n + d_{n-1}(\mathbf{p})s^{n-1} + \cdots\cdots + d_1(\mathbf{p})s + d_0(\mathbf{p})} \tag{2.10}$$

$$\mathbf{C}_i(s) = \frac{\beta_{it}s^t + \beta_{it-1}s^{t-1} + \cdots\cdots + \beta_{i1}s + \beta i0}{\alpha_t s^t + \alpha_{t-1}s^{t-1} + \cdots\cdots + \alpha_1 s + \alpha_0} \quad . \tag{2.11}$$

Then with

$$c(\mathbf{x}) := (\,\alpha_t \;\cdots\; \alpha_0 \;,\beta_{1t} \;\cdots\; \beta_{10}, \;\cdots, \;\beta_{mt} \;\cdots\; \beta_{m0}\,)^T := \mathbf{x} \tag{2.12}$$

as the controller parameter vector we get for the closed loop characteristic vector

$$\underbrace{\begin{pmatrix} d_n(\mathbf{p}) & 0 & \cdots & n_{1n}(\mathbf{p}) & 0 & \cdots & \cdots & n_{mn}(\mathbf{p}) & 0 \\ d_{n-1}(\mathbf{p}) & \ddots & \cdots & n_{1n-1}(\mathbf{p}) & \ddots & \cdots & \cdots & n_{mn-1}(\mathbf{p}) & \ddots \\ \vdots & \ddots & 0 & \vdots & \ddots & 0 & \cdots & \vdots & \ddots \\ \vdots & \ddots & d_n(\mathbf{p}) & \vdots & \ddots & n_{1n}(\mathbf{p}) & \cdots & \vdots & \ddots \\ d_0(\mathbf{p}) & \ddots & d_{n-1}(\mathbf{p}) & n_{10}(\mathbf{p}) & \ddots & n_{1n-1}(\mathbf{p}) & \cdots & n_{m0}(\mathbf{p}) & \ddots \\ 0 & \ddots & \vdots & 0 & \ddots & \vdots & \cdots & 0 & \ddots \\ \vdots & \ddots & d_0(\mathbf{p}) & \vdots & \ddots & n_{10}(\mathbf{p}) & \cdots & \vdots & \ddots \end{pmatrix}}_{\mathbf{M}(\mathbf{p})}$$

$$\underbrace{\begin{pmatrix} \alpha_t \\ \vdots \\ \alpha_0 \\ \beta_{1t} \\ \vdots \\ \beta_{10} \\ \vdots \\ \beta_{mt} \\ \vdots \\ \beta_{m0} \end{pmatrix}}_{c(\mathbf{x})} = \underbrace{\begin{pmatrix} \delta_{n+t} \\ \delta_{n+t-1} \\ \vdots \\ \vdots \\ \delta_1 \\ \delta_0 \end{pmatrix}}_{\delta}. \tag{2.13}$$

The multiinput single output case is the dual of the single input multioutput case with $\mathbf{G}(\mathbf{s})$ in (2.8) replaced by $\mathbf{G}(\mathbf{s})^T$ and $\mathbf{C}(\mathbf{s})$ in (2.9) replaced by $\mathbf{C}(\mathbf{s})^T$. The resulting equation for the characteristic vector is identical to (2.13).

2.3 Multiinput - Multioutput(MIMO) systems

Let $\mathbf{G}(s)$ denote the plant transfer function matrix and (A, B, C) a minimal realization. Let $(D(s), N(s))$ denote a left coprime factorization of $G(s)$ so that

$$\mathbf{G}(s) = C(sI - A)^{-1}B = D(s)^{-1}N(s). \tag{2.14}$$

Now let $\mathbf{C}(s)$ be a tth order proper feedback controller with minimal realization (A_c, B_c, C_c, D_c). Let

$$K_t := \begin{pmatrix} D_c & C_c \\ B_c & A_c \end{pmatrix} \tag{2.15}$$

$$A_t := \begin{pmatrix} A & 0 \\ 0 & 0 \end{pmatrix} \quad B_t := \begin{pmatrix} B & 0 \\ 0 & I_t \end{pmatrix} \quad C_t := \begin{pmatrix} C & 0 \\ 0 & I_t \end{pmatrix} \tag{2.16}$$

$$\mathbf{G}_t(s) := C_t(sI - A_t)^{-1}B_t \tag{2.17}$$

and note that the feedback system of Figure 2.1a has the same characteristic polynomial as that of the feedback system of Figure 2.1b.

Now

$$\begin{aligned} \mathbf{G}_t(s) : &= \begin{pmatrix} C(sI - A)^{-1}B & 0 \\ 0 & \frac{1}{s}I_t \end{pmatrix} \\ &= \underbrace{\begin{pmatrix} D(s) & 0 \\ 0 & sI_t \end{pmatrix}^{-1}}_{D_m(s)^{-1}} \underbrace{\begin{pmatrix} N(s) & 0 \\ 0 & I_p \end{pmatrix}}_{N_m(s)} \end{aligned} \tag{2.18}$$

is a left coprime factorization of $\mathbf{G}_t(s)$ and

$$K_t := \underbrace{K_t}_{N_c(s)} \underbrace{I^{-1}}_{D_c(s)^{-1}} \tag{2.19}$$

is a right coprime factorization of K_t.

Then

$$\delta(s) = \mathrm{Det}[D_m(s)D_c(s) + N_m(s)N_c(s)] \tag{2.20}$$

or

$$\delta(s) = \mathrm{Det}\left\{ \begin{pmatrix} D_m(s) & N_m(s) \end{pmatrix} \begin{pmatrix} D_c(s) \\ N_c(s) \end{pmatrix} \right\}.$$

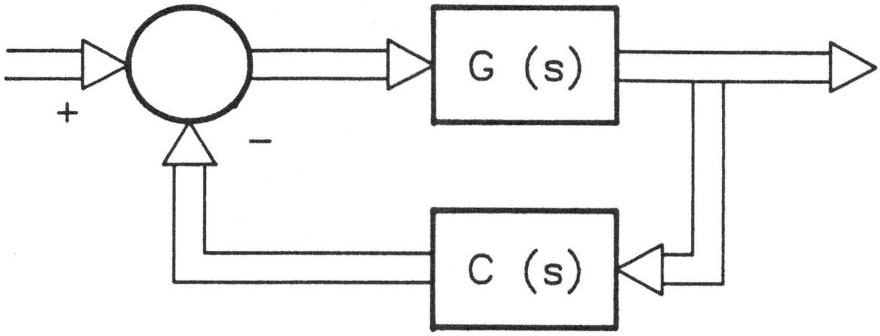

Feedback System

Figure 2.1a

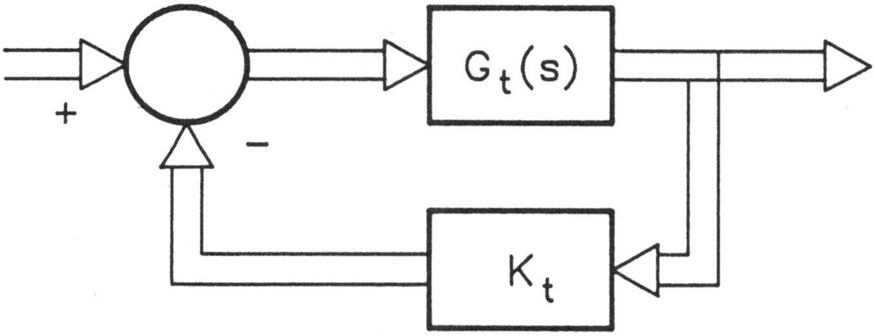

Equivalent Feedback System

Figure 2.1b

and from (2.18) and (2.19)

$$\delta(s) = \mathrm{Det}\underbrace{\begin{pmatrix} D(s) & 0 & \vdots & N(s) & 0 \\ 0 & sI_t & \vdots & 0 & I_t \end{pmatrix}}_{P(s)}\underbrace{\begin{pmatrix} I \\ K_t \end{pmatrix}}_{Q}$$

$$:= \mathrm{Det}[P(s)Q] \,. \tag{2.21}$$

Let the vector $\mathbf{x}$ equal an ordered list of entries of K_t. From the Binet-Cauchy formula [54], Det$[PQ]$ can be expanded as a sum of products of appropriate determinants of submatrices of P and Q. Let

$$P\begin{pmatrix} i_1 & i_2 & \cdots & i_p \\ j_1 & j_2 & \cdots & j_p \end{pmatrix} \tag{2.22}$$

denote the determinant of the submatrix of P formed by the rows $i_1, i_2, \cdots, i_p$ and the columns $j_1, j_2, \cdots, j_p$. Similar definitions apply to Q. Now

$$\begin{aligned} &\mathrm{Det}[P(s)Q] \\ &= \sum_{1 \le j_1 < \cdots < j_r \le n+m} P\begin{pmatrix} 1 & 2 & \cdots & n \\ j_1 & j_2 & \cdots & j_n \end{pmatrix} Q\begin{pmatrix} j_1 & j_2 & \cdots & j_n \\ 1 & 2 & \cdots & n \end{pmatrix} \\ &= \sum_{i=1}^{z} p_i(s) q_i \,. \end{aligned} \tag{2.23}$$

Therefore,

$$\delta(s) = \sum_{i=1}^{z} p_i(s) q_i, \qquad z = \binom{n+m}{n}. \tag{2.24}$$

Note that the polynomials $p_i(s)$ are functions only of the polynomials derived from the plant and the q_i are likewise funtions only of $\mathbf{x}$ derived from the controller. Therefore, if $\mathbf{p}$ denotes a vector of plant parameters, the coefficients of the $p_i(s)$ are functions only of $\mathbf{p}$. Similarly, the constants q_i are functions only of $\mathbf{x}$. It follows that the matrix $\mathbf{M(p)}$

consists of the coefficients of $p_i(s)$ and $c(\mathbf{x})$ consists of the q_i respectively. Therefore we have

$$\underline{\mathbf{M}(\mathbf{p})c(\mathbf{x}) = \delta.} \tag{2.25}$$

3. STABILITY MARGIN

Let $\mathbf{p}^0$ denote the nominal value of the plant parameter vector and $\mathbf{p}^0 + \Delta\mathbf{p}$ a perturbation. Then

$$\mathbf{M}(\mathbf{p}) := \mathbf{M}(\mathbf{p}^0 + \Delta\mathbf{p}) = \mathbf{M}(\mathbf{p}^0) + \Delta\mathbf{M}(\mathbf{p}^0, \Delta\mathbf{p}) \tag{3.1}$$

defines the perturbation of $\mathbf{M}(\mathbf{p}^0)$. Let $\mathbf{x}$ denote the controller parameter vector. Then the nominal characteristic vector is $\mathbf{M}(\mathbf{p}^0)c(\mathbf{x}) := \delta^0$. Under perturbations of $\mathbf{p}^0$ the characteristic vector δ^0 suffers perturbations given by

$$\mathbf{M}(\mathbf{p}^0 + \Delta\mathbf{p})c(\mathbf{x}) := \delta := \delta^0 + \Delta\delta. \tag{3.2}$$

In this setting, the problem of robust stabilization is to ensure that the characteristic vector δ remains strictly Hurwitz (i.e. the corresponding $\delta(s)$ has roots in the open left half plane) for the given class of perturbations denoted by $\{\Delta\mathbf{p}\} := \underline{\Delta}$. If, for example, the class of perturbations consists of perturbed ranges of parameter excursion,then

$$\underline{\Delta} := \{\Delta\mathbf{p} | -\gamma_i < \Delta\mathbf{p}_i < \epsilon_i, i = 1, \cdots, k\}. \tag{3.3}$$

Let $\rho(\delta^0)$ denote, as in Chapter 1, the radius of the stability hypersphere centered at δ^0 in the space δ, i.e.$\delta^0 + \Delta\delta$ is strictly Hurwitz for all $\|\Delta\delta\|_2 < \rho(\delta^0)$ and there exists $\overline{\Delta\delta}$ with $\|\overline{\Delta\delta}\|_2 = \rho(\delta^0)$ such that $\delta^0 + \overline{\Delta\delta}$ is not strictly Hurwitz. An algorithm for calculating

$\rho(\delta^0)$ is given later in this section. From (3.1) and (3.2) it follows that robust stability is achieved if

$$\|\Delta\mathbf{M}(\mathbf{p}^0,\Delta\mathbf{p})c(\mathbf{x})\| < \rho(\delta^0) \qquad \forall\Delta\mathbf{p}\in\underline{\Delta}\ . \tag{3.4}$$

Let $\|\cdot\|_F$ denote the Frobenius norm. The inequality (3.4) can be satisfied if

$$\|\Delta\mathbf{M}(\mathbf{p}^0,\Delta\mathbf{p})\|_F\|c(\mathbf{x})\|_2 < \rho(\delta^0) \qquad \forall\Delta\mathbf{p}\in\underline{\Delta} \tag{3.5}$$

or equivalently

$$\frac{\rho(\delta^0)}{\|c(\mathbf{x})\|_2} > \|\Delta\mathbf{M}(\mathbf{p}^0,\Delta\mathbf{p})\|_F \qquad \forall\Delta\mathbf{p}\in\underline{\Delta}. \tag{3.6}$$

Let

$$\sup_{\Delta\mathbf{p}\in\underline{\Delta}} \|\Delta\mathbf{M}(\mathbf{p}^0,\Delta\mathbf{p})\|_F := \beta. \tag{3.7}$$

We have now proved the following result.

<u>Theorem 3.1</u>

<u>Let $\mathbf{x}$ be a stabilizing controller when $\mathbf{p}=\mathbf{p}^0$. Then $\mathbf{x}$ stabilizes the closed loop system for all perturbations</u> $\Delta\mathbf{p}\in\underline{\Delta}$ <u>if</u>

$$\underline{\frac{\rho(\mathbf{M}(\mathbf{p}^0)c(\mathbf{x}))}{\|c(\mathbf{x})\|_2} > \beta}\ . \tag{3.8}$$

The formula (3.8) indicates that the controller $\mathbf{x}$ must not only enlarge the stability hypersphere,but do so with small "gain" $\|c(\mathbf{x})\|_2$.

The quantity

$$\mu(\mathbf{x}) := \frac{\rho(\delta^0)}{\|c(\mathbf{x})\|_2} = \frac{\rho(\mathbf{M}(\mathbf{p}^0)c(\mathbf{x}))}{\|c(\mathbf{x})\|_2} \tag{3.9}$$

is now proposed as a stability margin for the system with the given controller $\mathbf{x}$. The justification for using this quantity as a stability margin is apparent from (3.8) which shows that this margin serves as an upper bound on the level β of perturbations that can be tolerated in $\mathbf{p}^0$ with guaranteed stability.

Note that the Frobenius norm was used in (3.5) – (3.7) instead of the sharper estimate $\|\Delta \mathbf{M}(\mathbf{p}^0, \Delta \mathbf{p})\|_2$ because the supremum of the latter quantity for $\Delta \mathbf{p} \in \underline{\Delta}$ is much more difficult to evaluate. The quantity β is independent of the controller and can often be easily calculated as the example in Section 5 shows.

The above results hold in general. In the case of single input or single output plants a sharper result can be obtained using the results of Chapter 2 where the radius of the stability hypersphere was calculated in the space of transfer function coefficients of the plant.

Single Input or Single Output Case

Consider a single input or single output plant with parameter $\mathbf{p}$ entering the transfer function coefficients nonlinearly.

In this case the characteristic vector is

$$\mathrm{X}m(\mathbf{p}^0) = \delta^0 \tag{3.11}$$

where X is the matrix of controller transfer function coefficients and $m(\mathbf{p}^0)$ is the vector of plant transfer function coefficients, as in equation (2.3.8) in Chapter 2.

Let

$$m(\mathbf{p}) := m(\mathbf{p}^0) + \Delta m(\mathbf{p}^0, \Delta \mathbf{p}). \tag{3.12}$$

Write

$$m(\mathbf{p}^0) = \mathbf{y}^0 \tag{3.13}$$

$$m(\mathbf{p}) = \mathbf{y}^0 + \Delta\mathbf{y} \tag{3.14}$$

and with **x** fixed, consider the largest stability hypersphere in the space **y** centered at $\mathbf{y}^0$ with radius $\rho_x(\mathbf{y}^0) = \rho_x(m(\mathbf{p}^0))$. Then closed loop stability is preserved if

$$\|\Delta m(\mathbf{p}^0, \Delta\mathbf{p})\|_2 < \rho_x(m(\mathbf{p}^0)) \quad \forall \Delta\mathbf{p} \in \underline{\Delta} \ . \tag{3.15}$$

Write

$$\sup_{\Delta\mathbf{p}\in\underline{\Delta}} \|\Delta m(\mathbf{p}^0, \Delta\mathbf{p})\|_2 := \alpha. \tag{3.16}$$

Theorem 3.2

Let **x** stabilize the nominal system. Then the closed loop system remains stable for all perturbations $\Delta\mathbf{p} \in \underline{\Delta}$ if

$$\alpha < \rho_x(m(\mathbf{p}^0)) \ . \tag{3.17}$$

This theorem is an improvement over the previous one because of the following argument. Using the previous notation

$$\mathrm{X}(m(\mathbf{p}^0) + \Delta m(\mathbf{p}^0, \Delta\mathbf{p})) = \delta^0 + \Delta\delta \tag{3.18}$$

and

$$\mathrm{X}\Delta m(\mathbf{p}^0, \Delta\mathbf{p}) = \Delta\delta \ . \tag{3.19}$$

Now if

$$\|\mathrm{X}\Delta m(\mathbf{p}^0, \Delta\mathbf{p})\|_2 < \rho(\delta^0) \tag{3.20}$$

stability holds. A sufficient condition for this is:

$$\|\Delta m(\mathbf{p}^0, \Delta\mathbf{p})\|_2 < \frac{\rho(\delta^0)}{\|\mathrm{X}\|_2}. \tag{3.21}$$

Therefore if

$$\sup_{\Delta\mathbf{p}\in\underline{\Delta}} \|\Delta m(\mathbf{p}^0, \Delta\mathbf{p})\|_2 = \alpha \tag{3.22}$$

the condition given by Theorem 3.1 can be written as

$$\alpha < \frac{\rho(\delta^0)}{\|\mathrm{X}\|_2} \quad . \tag{3.23}$$

The condition (3.17) given by Theorem 3.2 is better because

$$\frac{\rho(\delta^0)}{\|\mathrm{X}\|_2} \leq \rho_x(m(\mathbf{p}^0)) \tag{3.24}$$

and therefore the bound on allowable perturbations given by this theorem is bigger. These arguments also lead to the following useful result.

Corollary 3.3

Under the conditions of Theorem 3.2 the controller stabilizes the system for all perturbations$\Delta\mathbf{p} \in \underline{\Delta}$ if

$$\alpha < \frac{\rho(\delta^0)}{\|\mathrm{X}\|_2} \quad . \tag{3.23}$$

MIMO Case

In this case the nominal characteristic equation can be written as

$$C(\mathbf{x})m(\mathbf{p}^0) = \delta^0 \tag{3.25}$$

and

$$C(\mathbf{x})m(\mathbf{p}^0) + C(\mathbf{x})\Delta m(\mathbf{p}^0, \Delta\mathbf{p}) = \delta^0 + \Delta\delta \ . \tag{3.26}$$

Now $\rho_x(m(\mathbf{p}^0))$ may be calculated as before with X in Theorem 2.3.1 replaced by $C(\mathbf{x})$ and we can state the corresponding condition

$$\alpha \leq \rho_{\mathbf{x}}(m(\mathbf{p}^0)) \tag{3.27}$$

for robust stability.

Calculation of $\rho(\delta)$

The main calculation in evaluating the stability margin $\mu(\mathbf{x})$ in Theorem 3.1 is the determinition of $\rho(\delta)$, the radius of the stability hypersphere. The quantity $\rho_x(m(\mathbf{p}^0))$ in Theorem 3.2 was calculated in Chapter 2 and although it is true that $\rho(\delta)$ can be calculated similarly, or by using the formulas given in [6], we give here an alternative calculation of $\rho(\delta)$ that may be simpler for computation.

Write

$$\begin{aligned}\delta(s) &= \delta_0 + \delta_1 s + \cdots + \delta_n s^n \\ &= \underbrace{\delta_e(s)}_{\text{even degree terms}} + \underbrace{\delta_0(s)}_{\text{odd degree terms}} .\end{aligned}$$

From Theorem 3.2 of Chapter 1,

$$\rho(\delta) = \min\{d_0,\ d_n,\ d\} \tag{3.28}$$

where

$$d := \inf_{0 \leq \omega \leq \infty} d(\omega) \tag{3.29}$$

and d_0, d_n and $d(\omega)$ are the Euclidean distances respectively between δ and the subspaces Δ_0, Δ_n and $\Delta(\omega)$ defined in equation (1.3.4) in Chapter 1 and corresponding to polynomials with roots at 0, ∞ and $\pm j\omega$. We can easily see that

$$d_0 = |\delta_0|,\ d_n = |\delta_n|.$$

The determination of $d(\omega)$ follows.

Proposition 3.3

The distance $d(\omega)$ between δ and $\Delta(\omega)$ is given by

i) $n = 2p$

$$d^2(\omega) = \frac{[\delta_e(j\omega)]^2}{1+\omega^2+\cdots+\omega^{4p}} + \frac{[\frac{\delta_o(j\omega)}{j\omega}]^2}{1+\omega^4+\cdots+\omega^{4(p-1)}}. \tag{3.30}$$

ii) $n = 2p+1$

$$d^2(\omega) = \frac{[\delta_e(j\omega)]^2 + [\frac{\delta_o(j\omega)}{j\omega}]^2}{1+\omega^4+\cdots+\omega^{4p}}. \tag{3.31}$$

Proof

Let $\mathcal{P}_n$ equal the set of polynomials of degree $\leq n$ and let $\mathcal{P}(\omega)$ denote the subset of $\mathcal{P}_n$ consisting of multiples of $(s^2+\omega^2)$. Note that $\mathcal{P}_n$ is a vector space of dimension $n+1, \mathcal{P}(\omega)$ is a vector space of dimension $n-1$, and

$$T = \{s^2+\omega^2,\ s^3+\omega^2 s,\ s^4+\omega^2 s^2, \cdots, s^n+\omega^2 s^{n-2}\}$$

is a basis for $\mathcal{P}(\omega)$. Define the inner product in $\mathcal{P}_n$:

$$< p,\ q >= \sum_{i=0}^{n} p_i\ q_i$$

and let P_r^ω denote the orthogonal projection on $\mathcal{P}(\omega)$. Then, it is clear that

$$d^2(\omega) = \|\delta - P_r^\omega(\delta)\|_2^2. \tag{3.32}$$

Write

$$\delta(s) = (s^2+\omega^2)q(s) + \underbrace{as+b}_{r(s)}$$

so that

$$\delta(j\omega) = \delta_e(j\omega) + \delta_0(j\omega) = aj\omega + b$$

and

$$a = \frac{\delta_0(j\omega)}{(j\omega)} \qquad b = \delta_e(j\omega). \tag{3.33}$$

Now

$$P_r^\omega[\delta(s)] = P_r^\omega[(s^2 + \omega^2)q(s) + r(s)]$$

$$= (s^2 + \omega^2)q(s) + P_r^\omega[r(s)]$$

so that

$$\|\delta - P_r^\omega(\delta)\|_2^2 = \|r - P_r^\omega(r)\|_2^2 \ . \tag{3.34}$$

i) $\underline{n = 2p}$: In this case

$$p_1(s) = 1 + \omega^2 s^2 + \omega^4 s^4 + \cdots + (-1)^p \omega^{2p} s^{2p} \tag{3.35}$$

$$p_2(s) = s - \omega^2 s^3 + \omega^4 s^5 + \cdots (-1)^{(p-1)} \omega^{2(p-1)} s^{2p-1} \tag{3.36}$$

form a basis for $\mathcal{P}^\perp(\omega)$. Thus

$$\|r - P_r^\omega(r)\|_2^2 = \frac{<r,\ p_1>^2}{\|p_1\|^2} + \frac{<r,\ p_2>^2}{\|p_2\|^2}$$

$$= \frac{b^2}{\|p_1\|^2} + \frac{a^2}{\|p_2\|^2} \tag{3.37}$$

and so the formula (3.30) follows from (3.33).

(ii) $\underline{n = 2p+1}$: In this case $p_1(s)$ in (3.35) is unchanged and

$$p_2(s) = s - \omega^2 s^2 + \cdots + (-1)^p \omega^{2p} s^{2p+1}$$

and (3.31) once again follows from (3.37) and (3.33). This completes the proof. ◊

The formulas (3.30) and (3.31) aid in the calcualtion of d in (3.29). Write

$$t = 1/\omega \tag{3.38}$$

and observe that

$$d = \min_{0 \leq \omega \leq \infty} d(\omega) = \min(d_1,\ d_2)$$

$$d_1 = \min_{0 \leq \omega \leq 1} d(\omega) \tag{3.39a}$$

$$d_2 = \min_{0 \leq t \leq 1} d(t). \tag{3.39b}$$

Thus the minimization over the infinite range in the determination of d can be replaced by the two minimizations over the finite ranges in (3.39). Proposition 3.3 given above and the formula (3.39) are reported in [69] and are due to Hervé Chapellat.

4. ROBUSTIFICATION PROCEDURE

Using the stability margins defined in the previous section an algorithm for controller design can be developed to iteratively upgrade the vector $\mathbf{x}$ of adjustable controller parameters to increase the stability margin. If $\mathbf{x}_k$ denotes the choice of controller at the kth iteration let the corresponding stability margin

$$\mu(\mathbf{x}_k) := \frac{\rho(\mathbf{M}(\mathbf{p}^0)c(\mathbf{x}_k))}{\|c(\mathbf{x}_k)\|_2}. \tag{4.1}$$

Our objective is to choose $\mathbf{x}_{k+1}$ so that

$$\mu(\mathbf{x}_{k+1}) > \mu(\mathbf{x}_k) \tag{4.2}$$

and $\mathbf{x}_{k+1}$ is stabilizing. From the equations

$$\mathbf{M}(\mathbf{p}^0)c(\mathbf{x}_{k+1}) = \delta^{k+1} \tag{4.3}$$

and

$$\mathbf{M}(\mathbf{p}^0)c(\mathbf{x}_k) = \delta_k \tag{4.4}$$

we see that if

$$\|c(\mathbf{x}_{k+1}) - c(\mathbf{x}_k)\| < \frac{\rho(\mathbf{M}(\mathbf{p}^0)c(\mathbf{x}_k))}{\|\mathbf{M}(\mathbf{p}^0)\|_2} \tag{4.5}$$

$\mathbf{x}_{k+1}$ is guaranteed to be stabilizing. The correction $\Delta\mathbf{x}_k = \mathbf{x}_{k+1} - \mathbf{x}_k$ is chosen via a gradient method based on numerical evaluation of the gradient of $\mu(\mathbf{x})$.

In the special case of single input or single output systems, the condition (3.34) given by Corollary 3.3 suggests that the stability margin can be taken to be the quantity

$$\mu(\mathbf{x}) = \frac{\rho(\mathrm{Xm}(\mathbf{p}^0))}{\|\mathrm{X}\|_2} \; .$$

When the perturbation ranges are given, β or α should first be determined (or estimated) for the prescribed controller order. Then the iterative improvement process can stop when $\mu(\mathbf{x}) > \beta$ or $\mu(\mathbf{x}) > \alpha$ is attained because then stability with respect to the perturbation range given is guaranteed.

Note that since the convexity of the function $\mu(\mathbf{x})$ is not established there is no guarantee that a global minimum will be found. A common procedure in such cases is to choose several initial guesses and to select the best answer. We next give an example of the robustification procedure.

5. EXAMPLE

The calculations of this chapter will be illustrated by considering the following multivariable system [23]:

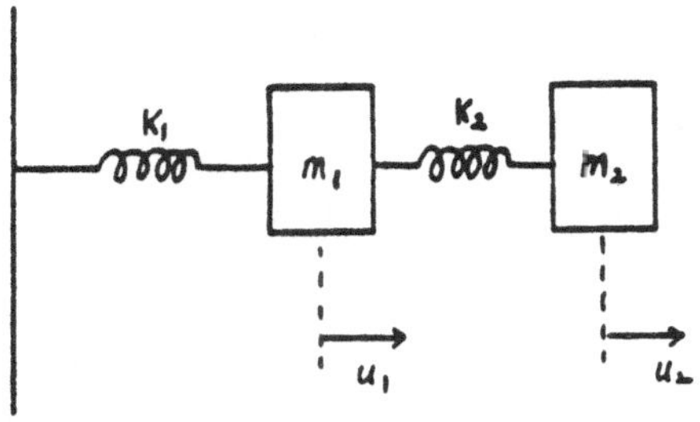

Figure 2.2 Two mass - two spring multivariable System

The transfer function of this marginally stable system is

$$\begin{pmatrix} y_1 \\ y_2 \end{pmatrix} = G(s) \begin{pmatrix} u_1 \\ u_2 \end{pmatrix}$$

$$G(s) =$$

$$\begin{pmatrix} \frac{s^2+2k_2}{\Delta(s)} & \frac{s^2+2k_2+k_1}{\Delta(s)} \\ \frac{s^3+2k_2s}{\Delta(s)} & \frac{s^3+(2k_2+k_1)s}{\Delta(s)} \end{pmatrix}$$

where

$$\Delta(s) = m_1m_2s^4 + (k_2m_2 + k_1m_2 + k_2m_1)s^2 + k_1k_2$$

We regard

$$\mathbf{p} := (\,m_1 \quad m_2 \quad k_1 \quad k_2\,)^T$$

as the physical parameter vector subject to perturbation. Now consider the 0th order controller

$$C(s) = \frac{1}{\gamma}\begin{pmatrix} \alpha_{11} & \alpha_{12} \\ \alpha_{21} & \alpha_{22} \end{pmatrix}$$

with the corresponding parameter vector

$$\mathbf{x} = (\,\gamma \quad \alpha_{11} \quad \alpha_{12} \quad \alpha_{21} \quad \alpha_{22}\,)^T.$$

The characteristic polynomial of the closed loop system becomes

$$\begin{aligned}\delta(s) &= \Delta(s)\det\{I + \mathrm{M}(s)C(s)\} \\ &= \gamma m_1 m_2 s^4 + (\alpha_{12} + \alpha_{22})s^3 + (\gamma k_2 m_2 + \gamma k_1 m_2 + \gamma k_2 m_1 + \alpha_{11} + \alpha_{21})s^2 \\ &+ (2k_2\alpha_{12} + 2k_2\alpha_{22} + k_1\alpha_{22})s \\ &+ (\gamma k_1 k_2 + 2k_2\alpha_{11} + 2k_2\alpha_{21} + k_1\alpha_{21})\,.\end{aligned}$$

If we write the coefficients of this polynomial in vector form by separating the plant parameters and the controller parameters, we have

$$\underbrace{\begin{pmatrix} m_1 m_2 & 0 & 0 & 0 & 0 \\ 0 & 1 & 1 & 0 & 0 \\ k_2 m_2 + k_1 m_2 + k_2 m_1 & 0 & 0 & 1 & 1 \\ 0 & 2k_2 & 2k_2 + k_1 & 0 & 0 \\ k_1 k_2 & 0 & 0 & 2k_2 & 2k_2 + k_1 \end{pmatrix}}_{\mathrm{M}(\mathbf{p})}$$

$$\underbrace{\begin{pmatrix} \gamma \\ \alpha_{12} \\ \alpha_{22} \\ \alpha_{11} \\ \alpha_{21} \end{pmatrix}}_{c(\mathbf{x})} = \underbrace{\begin{pmatrix} \delta_4 \\ \delta_3 \\ \delta_2 \\ \delta_1 \\ \delta_0 \end{pmatrix}}_{\delta}.$$

Now let $m_1 = 1, m_2 = 2, k_1 = 1, k_2 = 2$ with each element perturbing as follows

$$\mathbf{p}^0 := (1 \quad 2 \quad 1 \quad 2)^T$$

$$\underline{\Delta} := \{\Delta m_1, \Delta m_2, \Delta k_1, \Delta k_2 |$$

$$|\Delta m_1| \leq 0.01, |\Delta m_2| \leq 0.01,$$

$$|\Delta k_1| \leq 0.01, |\Delta k_2| \leq 0.01\} .$$

Then

$$\mathrm{M}(\mathbf{p}^0) = \begin{pmatrix} 2 & 0 & 0 & 0 & 0 \\ 0 & 1 & 1 & 0 & 0 \\ 8 & 0 & 0 & 1 & 1 \\ 0 & 4 & 5 & 0 & 0 \\ 2 & 0 & 0 & 4 & 5 \end{pmatrix}$$

and

$$\beta = \sup_{\Delta \mathbf{p} \in \underline{\Delta}} \|\Delta \mathrm{M}(\mathbf{p}^0, \Delta \mathbf{p})\|_F =$$

$$\sup_{\Delta p \in \underline{\Delta}} \left\| \begin{pmatrix} \Delta m_1 m_2 + m_1 \Delta m_2 + \Delta m_1 \Delta m_2 & 0 & 0 & 0 & 0 \\ 0 & 0 & 0 & 0 & 0 \\ \begin{matrix} \Delta k_2 m_2 + k_2 \Delta m_2 + \Delta k_2 \Delta m_2 \\ + \Delta k_1 m_2 + k_1 \Delta m_2 + \Delta k_1 \Delta m_2 \\ + \Delta k_2 m_1 + k_2 \Delta m_1 + \Delta k_2 \Delta m_1 \end{matrix} & 0 & 0 & 0 & 0 \\ 0 & 2\Delta k_2 & 2\Delta k_2 + \Delta k_1 & 0 & 0 \\ \Delta k_1 k_2 + k_1 \Delta k_2 + \Delta k_1 \Delta k_2 & 0 & 0 & 2\Delta k_2 & 2\Delta k_2 + \Delta k_1 \end{pmatrix} \right\|_F$$

$$= \left\| \begin{pmatrix} 0.0301 & 0 & 0 & 0 & 0 \\ 0 & 0 & 0 & 0 & 0 \\ 0.0903 & 0 & 0 & 0 & 0 \\ 0 & 0.02 & 0.03 & 0 & 0 \\ 0.0301 & 0 & 0 & 0.02 & 0.03 \end{pmatrix} \right\|_F = 0.112099 .$$

Choosing the initial stabilizing controller as

$$\mathbf{x}_0 = \begin{pmatrix} 2.2649934 \\ 13.898785 \\ -11.01622 \\ 7.4233423 \\ -6.212874 \end{pmatrix}$$

we get the roots of the resulting closed loop system to be

$$\begin{pmatrix} -6.5543886\mathrm{x}10^{-4} \pm j0.4126079 \\ -0.317509796 \pm j1.99883004 \end{pmatrix}$$

The corresponding radius of the stability hypersphere and stability margin are

$$\rho(\delta^0) = 0.0227866$$

$$\mu(\mathbf{x_0}) = 0.0011208.$$

Since $\mu(\mathbf{x_0}) < \beta$ the stability margin is inadequate and the initial choice of the stabilizing controller needs to be robustified. After 19 iterations of the robustification procedure of Section 4 we have a new controller

$$\mathbf{x}^* = \begin{pmatrix} 3.94658 \\ 15.8302 \\ -12.017 \\ 0.28319 \\ -1.2456 \end{pmatrix}.$$

The characteristic roots of the closed loop system are

$$\begin{pmatrix} -4.959486\text{x}10^{-2} \pm j0.30490336 \\ -0.190843011 \pm j1.925854956 \end{pmatrix}.$$

The corresponding stability radius and stability margin are

$$\rho(\delta^*) = 2.833679$$

$$\mu(\mathbf{x}^*) = 0.139547\,.$$

Since $\mu(\mathbf{x}^*) > \beta$, Theorem 3.1 shows that this controller guarantees stability for the given range of perturbations.

Remarks

Although the formulation presented in this chapter is completely general the results obtained are conservative in relation to the previous chapters. This conservatism stems from the use of the radius of the stability hyphersphere in coefficient space, $\rho(\delta)$, in calculating the stability margin. These results could be sharpened if larger stabilty regions could be determined in the parameter space. This is a difficult open problem in the general case which deserves much further study.

On the positive side the stability margins and robustification methods established here are computationally simple and provide some insight into the multivariable robust stabilization problem.

CHAPTER 5

STRUCTURED PERTURBATIONS IN STATE SPACE MODELS

1. INTRODUCTION

In the last three chapters, we concentrated on the transfer function description of the plant and derived conditions for robust stabilization based on analysis of the closed loop characteristic polynomial.

In this chapter, we consider situations where the plant description is given in the state space format. In such cases, the matrices that make up the state space model contain various physical parameters subject to perturbation and the robust stability and stabilization problems are solved most naturally in this setting. This is done in the present chapter. The problem formulation and the main results are given in the next section. This is followed by an example. The Appendix contains derivations of the gradient evaluations.

2. STABILITY MARGIN AND ROBUSTIFICATION

2.1 Problem Formulation

Assume now that the plant equations are derived from physical considerations in the state space form

$$\begin{aligned} \dot{x} &= Ax + Bu \\ y &= Cx \end{aligned} \tag{2.1}$$

and let the controller of order t be described by

$$\begin{aligned} \dot{x}_c &= A_c x_c + B_c y \\ u &= C_c x_c + D_c y. \end{aligned} \tag{2.2}$$

The closed loop system equations are

$$\begin{pmatrix} \dot{x} \\ \dot{x}_c \end{pmatrix} = \begin{pmatrix} A + BD_cC & BC_c \\ B_cC & A_c \end{pmatrix} \begin{pmatrix} x \\ x_c \end{pmatrix}$$
$$= \Big\{ \underbrace{\begin{pmatrix} A & 0 \\ 0 & 0_t \end{pmatrix}}_{A_t} + \underbrace{\begin{pmatrix} B & 0 \\ 0 & I_t \end{pmatrix}}_{B_t} \underbrace{\begin{pmatrix} D_c & C_c \\ B_c & A_c \end{pmatrix}}_{K_t} \underbrace{\begin{pmatrix} C & 0 \\ 0 & I_t \end{pmatrix}}_{C_t} \Big\} \begin{pmatrix} x \\ x_c \end{pmatrix}. \tag{2.3}$$

Now (2.2) is a stabilizing controller if and only if $A_t + B_tK_tC_t$ is stable. Since we will consider the compensator order to be fixed at each stage of the design process we drop the subscript t henceforth and consider the problem of robustification of $A + BKC$ by choice of K when the plant matrices are subject to perturbation.

Let $\mathbf{p} = (\, p_1 \quad p_2 \quad \cdots \quad p_r \,)$ denote a parameter vector consisting of physical parameters that enter the state space description linearly. This situation occurs frequently since the state equations are often written based on physical considerations. In any case, combinations of primary parameters can always be defined so that the resulting dependence of A, B, C on $\mathbf{p}$ is linear. We also assume that the nominal model (2.1) has been determined with the nominal value $\mathbf{p}^0$ of $\mathbf{p}$. This allows us to treat $\mathbf{p}$ purely as a perturbation with nominal value $\mathbf{p}^0 = 0$. Finally, since the perturbation enters at different locations we consider that $A + BKC$ perturbs to $A + BKC + \sum_{i=1}^{r} p_iE_i$ for given matrices E_i which prescribe the structure of the perturbation. For fixed K our problem is to determine the allowable perturbation in p_i that preserve stability.

2.1 Stability Margin

We now state a result that calculates the radius of a spherical stability region in the parameter space $\mathbf{p} \in R^r$ when the controller is given. This result will also be a useful step in the robustification procedure to be developed.

Let the nominal asymptotically stable system be

$$\dot{x}(t) = Mx(t) = (A + BKC)x(t) \tag{2.4}$$

and the perturbed equation be

$$\dot{x}(t) = (M + \sum_{i=1}^{r} p_i E_i)x(t) \tag{2.5}$$

where the $p_i, i = 1, \cdots, r$ are perturbations of parameters of interest and the $E_i, i = 1, \cdots, r$ are matrices determined by the structure of the parameter perturbations. Let $Q > 0$ be a positive definite symmetric matrix and let P denote the unique positive definite symmetric solution of

$$M^T P + PM + Q = 0 \, . \tag{2.6}$$

Theorem 2.1

The system (2.5) is stable for all p_i satisfying

$$\sum_{i=1}^{r} |p_i|^2 < \frac{\sigma_{min}^2(Q)}{\sum_{i=1}^{r} \mu_i^2} \tag{2.7}$$

where

$$\mu_i := \|E_i^T P + PE_i\|_2 \, .$$

Proof

Under the assumption that M is asymptotically stable with the stabilizing controller K, choose as the Lyapunov function

$$V(x) = x^T Px \tag{2.8}$$

where P is the symmetric positive definite solution of (2.6). Since M is an asymptotically stable matrix, the existence of such a P is guaranteed by Lyapunov's theorem. Note that $V(x) > 0$ for all $x \neq 0$ and $V(x) \longrightarrow \infty$ as $\|x\| \longrightarrow \infty$. We require $\dot{V}(x) \leq 0$ for the stability of (2.5). Differentiating (2.8) with respect to x along solutions of (2.5) yields

$$\begin{aligned} \dot{V}(x) &= \dot{x}^T P x + x^T P \dot{x} \\ &= x^T(M^T P + PM)x + x^T(\sum p_i E_i^T P + \sum p_i P E_i)x. \end{aligned} \tag{2.9}$$

Substituting (2.6) into (2.9) we have

$$\dot{V}(x) = -x^T Q x + x^T(\sum_{i=1}^{r} p_i E_i^T P + \sum_{i=1}^{r} p_i P E_i)x. \tag{2.10}$$

The stability requirement $\dot{V}(x) \leq 0$ is equivalent to

$$x^T(\sum_{i=1}^{r} p_i E_i^T P + \sum_{i=1}^{r} p_i P E_i)x \leq x^T Q x. \tag{2.11}$$

Using the Rayleigh principle [54],

$$\sigma_{min}(Q) \leq \frac{x^T Q x}{x^T x} \leq \sigma_{max}(Q) \qquad \forall x \neq 0 \tag{2.12}$$

we have

$$\sigma_{min}(Q) x^T x \leq x^T Q x. \tag{2.13}$$

Thus equation (2.11) is satisfied if

$$x^T(\sum_{i=1}^{r} p_i E_i^T P + \sum_{i=1}^{r} p_i P E_i)x \leq \sigma_{min}(Q) x^T x. \tag{2.14}$$

Since

$$|x^T(\sum_{i=1}^{r} p_i E_i^T P + \sum_{i=1}^{r} p_i P E_i)x|$$

$$\leq \|x^T\|_2 \|(\sum_{i=1}^{r} p_i E_i^T P + \sum_{i=1}^{r} p_i P E_i)\|_2 \|x\|_2$$

$$\leq \|x\|_2^2 (\sum_{i=1}^{r} |p_i| \|E_i^T P + P E_i\|_2) \tag{2.15}$$

(2.14) is satisfied if

$$\sum_{i=1}^{r} (|p_i| \|E_i^T P + P E_i\|_2) \leq \sigma_{min}(Q). \tag{2.16}$$

Let $\mu_i := \|E_i^T P + P E_i\|_2 = \sigma_{max}(E_i^T P + P E_i)$. Then (2.16) can be rewritten as

$$\sum_{i=1}^{r} (|p_i| \|E_i^T P + P E_i\|_2)$$

$$= \underbrace{\begin{pmatrix} |p_1| & |p_2| & \cdots & \cdots & |p_r| \end{pmatrix}}_{\underline{p}} \underbrace{\begin{pmatrix} \mu_1 \\ \mu_2 \\ \vdots \\ \vdots \\ \mu_r \end{pmatrix}}_{\underline{\mu}} \leq \sigma_{min}(Q) \tag{2.17}$$

which is satisfied if

$$\|\underline{p}\ \underline{\mu}\|_2^2 \leq \|\underline{p}\|_2^2 \|\underline{\mu}\|_2^2 \leq \sigma_{min}^2(Q). \tag{2.18}$$

Using the fact that

$$\|\underline{p}\|_2^2 = \sum_{i=1}^{r} |p_i|^2 \tag{2.19}$$

$$\|\underline{\mu}\|_2^2 = \sum_{i=1}^{r} \mu_i^2 \tag{2.20}$$

we obtain

$$\sum_{i=1}^{r} |p_i|^2 \leq \frac{\sigma_{min}^2(Q)}{\sum_{i=1}^{r} \mu_i^2}. \quad \diamond \tag{2.21}$$

This theorem determines for the given stabilizing controller K, the quantity

$$\rho(K, Q) := \sqrt{\frac{\sigma_{min}^2(Q)}{\sum_{i=1}^{r} \mu_i^2}} = \sqrt{\frac{\sigma_{min}^2(Q)}{\sum_{i=1}^{r} \|E_i^T P + P E_i\|_2^2}} \tag{2.22}$$

which determines the range of perturbations for which stability is guaranteed. This therefore is the radius of a stability hypersphere in parameter space.

2.3 Robustification Procedure

Using the index obtained in the previous subsection, we now give an iterative design procedure to obtain the optimal controller K^* so that (2.22) is as large as possible. For a given K the largest stability hypersphere we can obtain is

$$\max_Q \rho^2(K,Q) = \max_Q \frac{\sigma^2_{min}(Q)}{\sum_{i=1}^r \mu_i^2} \,. \tag{2.23}$$

Therefore the problem of designing a robust controller with respect to structured parameter perturbations can be formulated as:

Find K to maximize (2.23), i.e.

$$\max_K\{\max_Q \rho^2(K,Q)\} = \max_K\{\max_Q \frac{\sigma^2_{min}(Q)}{\sum_{i=1}^r \mu_i^2}\} \tag{2.24}$$

subject to

$$\sigma(A+BKC) \subset C^-.$$

Equivalently

$$\max_{K,Q} \rho^2(K,Q) = \max_{K,Q} \frac{\sigma^2_{min}(Q)}{\sum_{i=1}^r \mu_i^2} \tag{2.25}$$

subject to

$$\sigma(A+BKC) \subset C^-.$$

The following constrainted optimization problem is therefore constructed: Given (A,B,C) find a stabilizing controller K and a real matrix L to minimize J given below

$$\min_{K,L} J := \min_{K,L} \frac{\sum_{i=1}^r \|E_i^T P + PE_i\|_2^2}{\sigma^2_{min}(L^T L)} \tag{2.26}$$

subject to

$$(A + BKC)^T P + P(A + BKC) = -L^T L \tag{2.27a}$$

and

$$J_c := \max_{\lambda \in \sigma(A+BKC)} \text{Real}(\lambda) < 0. \tag{2.27b}$$

Note that the positive definite matrix Q has been replaced without loss of generality by $L^T L$. For any square full rank matrix L, $L^T L$ is positive definite symmetric. This replacement also reduces computational complexity.

In order to implement a gradient based descent procedure we derive the gradient of (2.27) with respect to K and L. Before we state this result consider a slightly more general class of perturbations, i.e.

$$A = A_0 + \sum_{i=1}^{r} p_i A_i, \qquad B = B_0 + \sum_{i=1}^{r} p_i B_i \tag{2.28}$$

Then we get

$$M = A_0 + B_0 KC \quad \text{and} \quad E_i = A_i + B_i KC. \tag{2.29}$$

Theorem 2.2

Let J be defined as in (2.26) and let (2.26) – (2.29) hold. Then

(a)

$$\frac{\partial J}{\partial L} = \frac{2}{\sigma_{min}^3(L^T L)} L\{\sigma_{min}(L^T L)V^T - \sum_{i=1}^{r} \sigma_{max}^2 (E_i^T P + PE_i)(u_m v_m^T + v_m u_m^T)\} \tag{2.30}$$

where V satisfies

$$(A_0 + B_0 KC)V + V(A_0 + B_0 KC)^T =$$

$$-\sum_{i=1}^{r}\sigma_{max}(E_i^TP+PE_i)\{E_i(u_{ai}v_{ai}^T+v_{ai}u_{ai}^T)+(u_{ai}v_{ai}^T+v_{ai}u_{ai}^T)E_i^T\} \tag{2.31}$$

where v_{ai} and u_{ai} are left and right singluar vectors corresponding to $\sigma_{max}(E_i^TP+PE_i)$, and v_m and u_m are left and right singluar vectors corresponding to $\sigma_{min}(L^TL)$.

(b)

$$\frac{\partial J}{\partial K}=\frac{2}{\sigma_{min}^2(L^TL)}$$

$$\{\sum_{i=1}^{r}\sigma_{max}(E_i^TP+PE_i)B_i^TP(v_{ai}u_{ai}^T+u_{ai}v_{ai}^T)+B^TP^TV^T\}C^T \tag{2.32}$$

(c)

$$\frac{\partial J_c}{\partial K_{ij}}=\text{Real}\{\frac{v^TB_0(\frac{\partial K}{\partial K_{ij}})Cw}{v^Tw}\} \tag{2.33}$$

where v and w are the left and right eigenvectors of (A_0+B_0KC) corresponding to λ_{max} the eigenvalue with $\max\{\text{Real}(\lambda)\}$.

The proof of this theorem is given in the Appendix.

An algorithm for enlarging the radius of the stability hypersphere $\rho(K,Q)$ by iterating on (K,Q) can be devised using these gradients. Such an algorithm has been implemented using the Harwell optimization package [48]. The iterations stop when

$$\rho(K_i,L_i)>\max||\Delta p||_2$$

is attained. Since little is known about the geometry of the function $\rho(K,Q)$ this procedure does not guarantee that a minimumim is attained but is nevertheless useful.

3. EXAMPLE

As an example we again consider the VTOL helicopter [53] considered in Example 3, Chapter 3. The linearized dynamic equation of the VTOL helicopter is shown below:

$$\frac{d}{dt}\begin{pmatrix} x_1 \\ x_2 \\ x_3 \\ x_4 \end{pmatrix} = \begin{pmatrix} -0.0366 & 0.0271 & 0.0188 & -0.4555 \\ 0.0482 & -1.010 & 0.0024 & -4.0208 \\ 0.1002 & p_1 & -0.707 & p_2 \\ 0 & 0 & 1 & 0 \end{pmatrix}\begin{pmatrix} x_1 \\ x_2 \\ x_3 \\ x_4 \end{pmatrix}$$

$$\begin{pmatrix} 0.4422 & 0.1761 \\ p_3 & -7.5922 \\ -5.52 & 4.49 \\ 0 & 0 \end{pmatrix}\begin{pmatrix} u_1 \\ u_2 \end{pmatrix}$$

$$y = (0 \quad 1 \quad 0 \quad 0)\mathbf{x} = x_2 \; .$$

The most significant changes take place in the elements p_1, p_2 and p_3. The following bounds on the parameters are given in [53]:

$$\begin{aligned} p_1 &= 0.3681 + \Delta p_1 \qquad |\Delta p_1| \leq 0.05 \\ p_2 &= 1.4200 + \Delta p_2 \qquad |\Delta p_2| \leq 0.01 \\ p_3 &= 3.5446 + \Delta p_3 \qquad |\Delta p_3| \leq 0.04 \end{aligned} \tag{3.1}$$

Now we compute

$$\max \|\Delta p\|_2 = 0.0648.$$

The eigenvalues of the open loop unstable plant are

$$\lambda(A) = \begin{pmatrix} 0.27579 \pm j0.25758 \\ -0.2325 \\ -2.072667 \end{pmatrix}.$$

A nominal stabilizing controller is given by

$$K_0 = \begin{pmatrix} -1.63522 \\ 1.582236 \end{pmatrix}.$$

In Chapter 7 we develop an algorithm to obtain such a stabilizing controller for a given plant. Starting with this nominal stabilizing controller we performed the robustification procedure of Section 2. For this step we took the initial value

$$L_0 = \begin{pmatrix} 1 & 0 & -0.5 & 0.06 \\ 0.5 & 1 & -0.03 & 0 \\ -0.1 & 0.4 & 1 & 0.14 \\ 0.2 & 0.6 & -0.13 & 1.5 \end{pmatrix}.$$

The nominal values gave the stability margin

$$\rho_0 = 0.02712 < 0.0648 = \|\Delta p\|_2$$

which does not satisfy the requirement (2.7). After 26 iterations of the robustification procedure we have

$$\rho^* = 0.12947 > 0.0648 = \|\Delta p\|_2 \tag{3.2}$$

which does satisfy the requirement of Theorem 2.1. The robust stabilizing 0th order controller computed is

$$K^* = \begin{pmatrix} -0.99633989 \\ 1.801833665 \end{pmatrix}$$

and the corresponding optimal L^*, P^* and the closed loop eigenvalues are

$$L^* = \begin{pmatrix} 0.51243 & 0.02871 & -0.1326 & 0.05889 \\ -0.0004 & 0.39582 & -0.0721 & -0.3504 \\ 0.12938 & 0.08042 & 0.51089 & -0.0145 \\ -0.0715 & 0.34789 & -0.0253 & 0.39751 \end{pmatrix}$$

$$P^* = \begin{pmatrix} 2.00394 & -0.3894 & -0.5001 & -0.4922 \\ -0.3894 & 0.36491 & 0.46352 & 0.19652 \\ -0.5001 & 0.46352 & 0.61151 & 0.29841 \\ -0.4922 & 0.19652 & 0.29841 & 0.98734 \end{pmatrix}$$

$$\lambda(A + BK^*C) = \begin{pmatrix} -18.396295 \\ -0.247592 \pm j1.2501375 \\ -0.0736273 \end{pmatrix}.$$

The robust controller guarantees stability for the class of perturbations given. It can be seen from (3.2) that the radius of the stability sphere is big enough that the feedback system with this controller remains stable even when the "worst" perturbation from the class (3.1) amplified by a factor of two is injected into the closed loop.

APPENDIX

Proof of Theorem 3.2

Let

$$J := \frac{\sum_{i=1}^{r} \sigma_{max}^2(E_i^T P + PE_i)}{\sigma_{min}^2(L^T L)} = \frac{\sum_{i=1}^{r} \text{Trace}\{\sigma_{max}^2(E_i^T P + PE_i)\}}{\text{Trace}\{\sigma_{min}^2(L^T L)\}} \ . \tag{A.1}$$

Then

$$\Delta J = \frac{1}{\sigma_{min}^4(L^T L)} \{\sum_{i=1}^{r} 2\sigma_{max}(E_i^T P + PE_i)\sigma_{min}^2(L^T L)\text{Trace}(\Delta\sigma_{max}(E_i^T P + PE_i)) - \sum_{i=1}^{r} 2\sigma_{max}^2(E_i^T P + PE_i)\sigma_{min}(L^T L)\text{Trace}(\Delta\sigma_{min}(L^T L))\} \ . \tag{A.2}$$

Here we note that

$$\Delta\sigma_{max}(E_i^T P + PE_i) = v_{ai}u_{ai}^T \Delta(E_i^T P + PE_i) = v_{ai}u_{ai}^T(E_i^T \Delta P + \Delta P E_i) \tag{A.3}$$

where v_{ai} and u_{ai} are right and left singular vectors corresponding to $\sigma_{max}(E_i^T P + PE_i)$, respectively and

$$\Delta\sigma_{min}(L^T L) = v_m u_m^T \Delta(L^T L) = v_m u_m^T(\Delta L^T L + L^T \Delta L) \tag{A.4}$$

where v_m and u_m are right and left singular vectors corresponding to $\sigma_{min}(L^T L)$, respectively.

(a) Calculation of $\frac{\partial J}{\partial L}$

Define

$$M := A + BKC \tag{A.5}$$

Perturbing the Lyapunov equation (2.26) with respect to L and rejecting second order terms, we have

$$M^T \Delta P + \Delta P M = -(\Delta L^T L + L^T \Delta L) \ . \tag{A.6}$$

From [61], the solution of the equation $(A.6)$ is of the form

$$\Delta P=\sum_{j=1}^{n}\sum_{k=1}^{n}\gamma_{jk}(M^T)^{j-1}(\Delta L^TL+L^T\Delta L)M^{k-1}. \tag{A.7}$$

Substituting $(A.7)$ into $(A.3)$ leads to

$$\text{Trace}(\Delta\sigma_{max}(E_i^TP+PE_i))=$$

$$\text{Trace}(v_{ai}u_{ai}^T(E_i^T\Delta P+\Delta PE_i))=$$

$$\text{Trace}\{\sum_j\sum_k\gamma_{jk}v_{ai}u_{ai}^TE_i^T(M^T)^j(\Delta L^TL+L^T\Delta L)M^k$$

$$+\sum_j\sum_k\gamma_{jk}v_{ai}u_{ai}^T(M^T)^j(\Delta L^TL+L^T\Delta L)M^kE_i\}=$$

$$\text{Trace}\{\sum_j\sum_k\gamma_{jk}M^j(E_i(u_{ai}v_{ai}^T+v_{ai}u_{ai}^T)+(u_{ai}v_{ai}^T+v_{ai}u_{ai}^T)E_i^T)(M^T)^kL^T\Delta L\} \tag{A.8}$$

and

$$\text{Trace}(\Delta\sigma_{min}(L^TL))=Trace(v_mu_m^T\Delta L^TL+v_mu_m^TL^T\Delta L)$$

$$=\text{Trace}(u_mv_m^TL^T\Delta L+v_mu_m^TL^T\Delta L) \tag{A.9}$$

$$=\text{Trace}\{(u_mv_m^T+v_mu_m^T)L^T\Delta L\}$$

Substituting $(A.8)$ and $(A.9)$ into $(A.2)$, we have

$$\Delta J=\frac{1}{\sigma_{min}^4(L^TL)}\{\sum_{i=1}^{r}\text{Trace}\{2\sigma_{max}(E_i^TP+PE_i)\sigma_{min}^2(L^TL)$$

$$\sum_j\sum_k\gamma_{jk}M^j(E_i(u_{ai}v_{ai}^T+v_{ai}u_{ai}^T)+(u_{ai}v_{ai}^T+v_{ai}u_{ai}^T)E_i^T)(M^T)^kL^T\Delta L\}$$

$$-\sum_{i=1}^{r}\text{Trace}\{2\sigma_{max}^2(E_i^TP+PE_i)\sigma_{min}(L^TL)(u_mv_m^T+v_mu_m^T)L^T\Delta L\}\}. \tag{A.10}$$

Now we have

$$\Delta J=\frac{2}{\sigma_{min}^3(L^TL)}\{\text{Trace}\{\sigma_{min}(L^TL)$$

$$\underbrace{\sum_{j,k}\gamma_{jk}M^j\sum_{i=1}^{r}\sigma_{max}(E_i^TP+PE_i)(E_i(u_{ai}v_{ai}^T+v_{ai}u_{ai}^T)+(u_{ai}v_{ai}^T+v_{ai}u_{ai}^T)E_i^T)(M^T)^k}_{V}$$

$$-\text{Trace}\{\sum_{i=1}^{r}\sigma_{max}^2(E_i^TP+PE_i)(u_mv_m^T+v_mu_m^T)\}\}L^T\Delta L$$

$$=\frac{2}{\sigma_{min}^3(L^TL)}$$

$$\text{Trace}\{\sigma_{min}(L^TL)V-\sum_{i=1}^{r}\sigma_{max}^2(E_i^TP+PE_i)(u_mv_m^T+v_mu_m^T)\}L^T\Delta L\ . \qquad (A.11)$$

Therefore,

$$\frac{\partial J}{\partial L}=\frac{2}{\sigma_{min}^3(L^TL)}L\{\sigma_{min}(L^TL)V^T-\sum_{i=1}^{r}\sigma_{max}^2(E_i^TP+PE_i)(u_mv_m^T+v_mu_m^T)\} \qquad (A.12)$$

where V satisfies

$$(A+BKC)V+V(A+BKC)^T=$$

$$-\sum_{i=1}^{r}\sigma_{max}(E_i^TP+PE_i)\{E_i(u_{ai}v_{ai}^T+v_{ai}u_{ai}^T)+(u_{ai}v_{ai}^T+v_{ai}u_{ai}^T)E_i^T\} \qquad (A.13)$$

as claimed.

(b) Calculation of $\frac{\partial J}{\partial K}$

Consider perturbations in $(A+BKC)$ and let

$$A:=A_0+\sum_{i=1}^{r}p_iA_i \qquad (A.14)$$

$$B:=B_0+\sum_{i=1}^{r}p_iB_i \qquad (A.15)$$

where A_i and B_i are given matrices. Therefore

$$p_iE_i:=p_i(A_i+B_iKC) \qquad (A.16)$$

and E_i is a function of K. Thus

$$\Delta E_i = B_i \Delta K C. \tag{A.17}$$

Now let

$$M := A_0 + B_0 K C. \tag{A.18}$$

Perturbing the Lyapunov equation with respect to K and rejecting second order terms, we have

$$M^T \Delta P + \Delta P M = -(C^T \Delta K^T B_0^T P + P B_0 \Delta K C) \tag{A.19}$$

$$\Delta P = \sum_j \sum_k \gamma_{jk} (M^T)^j (C^T \Delta K^T B_0^T P + P B_0 \Delta K C) M^k. \tag{A.20}$$

Now, differentiating $(A.1)$ with respect to K,

$$\begin{aligned} \Delta J &= \frac{1}{\sigma_{min}^2(L^T L)} \text{Trace}\{\sum_{i=1}^r 2\sigma_{max}(E_i^T P + P E_i) \Delta \sigma_{max}(E_i^T P + P E_i)\} \\ &= \frac{2}{\sigma_{min}^2(L^T L)} \sum_{i=1}^r \sigma_{max}(E_i^T P + P E_i) \text{Trace}\{\Delta \sigma_{max}(E_i^T P + P E_i)\} . \end{aligned} \tag{A.21}$$

Hence

$$\begin{aligned} \text{Trace}\{\Delta \sigma_{max}(E_i^T P + P E_i)\} &= \text{Trace}\{v_{ai} u_{ai}^T \Delta (E_i^T P + P E_i)\} \\ &= \text{Trace}\{v_{ai} u_{ai}^T (\Delta E_i^T P + E_i^T \Delta P + \Delta P E_i + P \Delta E_i)\} \\ &= \text{Trace}\{(u_{ai} v_{ai}^T + v_{ai} u_{ai}^T) P \Delta E_i + (v_{ai} u_{ai}^T E_i^T + E_i v_{ai} u_{ai}^T) \Delta P\} . \end{aligned} \tag{A.22}$$

Substituting $(A.17)$ and $(A.20)$ into $(A.22)$ we have

$$\text{Trace}\{\Delta \sigma(E_i^T P + P E_i)\} = \text{Trace}\{C(u_{ai} v_{ai}^T + v_{ai} u_{ai}^T) P B_i \Delta K + C$$

$$\sum_{j,k} \gamma_{jk} M^j ((u_{ai} v_{ai}^T + v_{ai} u_{ai}^T) E_i^T + E_i (u_{ai} v_{ai}^T + v_{ai} u_{ai}^T)) (M^T)^k P B \Delta K\}. \tag{A.23}$$

Substituting $(A.23)$ into $(A.21)$ leads to

$$\Delta J = \frac{2}{\sigma_{min}^2(L^TL)} C\text{Trace}\{\sum_{i=1}^{r} \sigma_{max}(E_i^TP + PE_i)(u_{ai}v_{ai}^T + v_{ai}u_{ai}^T)PB_i +$$

$$\underbrace{\sum_{j,k} M^j \sum_{i=1}^{r} \sigma_{max}(E_i^TP + PE_i)((u_{ai}v_{ai}^T + v_{ai}u_{ai}^T)E_i^T + E_i(u_{ai}v_{ai}^T + v_{ai}u_{ai}))(M^T)^k}_{V}$$

$$PB\}\Delta K \qquad (A.24)$$

$$= \frac{2}{\sigma_{min}^2(L^TL)} C\text{Trace}$$

$$\{\sum_{i=1}^{r} \sigma_{max}(E_i^TP + PE_i)(u_{ai}v_{ai}^T + v_{ai}u_{ai}^T)PB_i + VPB\}\Delta K \ . \qquad (A.25)$$

Therefore,

$$\frac{\partial J}{\partial K} =$$

$$\frac{2}{\sigma_{min}^2(L^TL)}\{\sum_{i=1}^{r} \sigma_{max}(E_i^TP + PE_i)B_i^TP(v_{ai}u_{ai}^T + u_{ai}v_{ai}^T) + B^TPV^T\}C^T \qquad (A.26)$$

where V satisfies $(A.13)$ and $E_i = A_i + B_iKC$ for $i = 1, \cdots, r$.

(c)

The gradients of the constraint function Real $\{\lambda_{max}(A + BKC)\} < 0$ with respect to L and K are:

$$J_c := Real\{\lambda_{max}(A_0 + B_0KC)\} \qquad (A.27)$$

$$\frac{\partial J_c}{\partial L} = 0 \qquad (A.28)$$

$$\frac{\partial J_c}{\partial K_{ij}} = \text{Real}\{\frac{v^TB_0(\frac{\partial K}{\partial K_{ij}})Cw}{v^Tw}\} \qquad (A.29)$$

with v and w are the corresponding left and right eigenvectors of $\lambda_{max}(A_0 + B_0KC)$.

CHAPTER 6

STABILIZATION WITH FIXED ORDER CONTROLLERS

1. INTRODUCTION

The theory of the previous chapters assumes, as a starting point, that a nominal stabilizing controller has been found. This chapter and the next deal with the problem of finding such a stabilizing compensator. When the dynamic order of the controller is fixed a priori this is an unsolved problem. Existing solutions to the regulator problem can only generate controllers that are of high enough order that arbitrary pole placement becomes possible. This includes the LQG theory [55], observed state feedback [56] and arbitrary pole placement approaches [57],[58]. Controllers that are robust with respect to unstructured perturbations evidently suffer from the same difficulty of high order (see the examples given in [2]). We also mention that adaptive control theory is notorious for producing high order solutions.

It is certainly essential in practice, to have low order solutions to the stabilization problem. This requirement arises because the controller must eventually carry out several functions such as tracking, disturbance rejection, desensitization against parameter variations, provide good transient response, small steady state error, prevent various signals from saturating etc., in addition to the basic task of stabilization. Many of these requirements are in conflict with each other in ways that cannot be handled analytically, and the only recourse left to the designer is to iteratively redesign the controller using adhoc methods and graphical displays until a satisfactory solution is obtained. This redesign must be carried out in the parameter space of the stabilizing controller. If the basic stabilizing controller order is high, so is the dimension of this parameter space and the subsequent design procedure can become unwieldy. From this perspective, the high order of controllers produced by "modern" control theory is one of the severest limitations of this theory.

We attempt to alleviate this problem by presenting, in this chapter, some methods in the transfer function domain, for designing low order stabilizing controllers. These results were originally reported in [47]. The formulation presented in this chapter deals with stabilization of a system with a controller of prescribed order. The results obtained can be used to find low order solutions by successively updating the prescribed order until a solution is found. This formulation allows us to obtain a lower bound on the order of a stabilizing controller by means of a classical result of linear programming, known as Gordan's Theorem [59]. Each stabilizing controller is found to correspond to a Hurwitz vector that lies in a linear subspace determined exclusively by the plant transfer function parameters, and the controller order. An algorithm for selecting Hurwitz vectors is described, that iteratively increases the radius of the largest stability hypersphere and reduces the distance to this linear subspace. A stabilizing controller of the prescribed order t is found when this largest stability hypersphere intersects the subspace.

2. NECESSARY CONDITIONS USING LINEAR PROGRAMMING

Consider the problem of stabilizing the fixed q^{th} order plant with transfer function G(s) with a controller $C(s)$of dynamic order t. The class of all controllers of order t can be parameterized by a real vector $\mathbf{x}$ of fixed dimension as in Chapters 2 and 4. For single input or single output plants $\mathbf{x}$ can consist of the list of transfer function coefficients of $C(s)$ as in equation (4.2.13). In multiinput multioutput systems $\mathbf{x}$ can consist of the list of entries of the state space representation of a tth order dynamic system with input y and output u as in equation (4.2.25). Let $n := t + q$ denote the dynamic order of the closed loop system and

$$\delta(s) = \sum_{i=0}^{n} \delta_i(\mathbf{x}) s^i \tag{2.1}$$

the characteristic polynomial of the closed loop system. As before let

$$\delta(\mathbf{x}) = [\delta_0(\mathbf{x}), \delta_1(\mathbf{x}), \cdots\cdots, \delta_n(\mathbf{x})^{\mathrm{T}}] \tag{2.2}$$

denote the characteristic vector, and let

$$H_n := [\delta | \delta \epsilon R^{n+1}, \delta(s) \text{ is of degree n and Hurwitz}] \,. \tag{2.3}$$

As shown in Chapter 2 and Chapter 4 we have

$$\mathbf{M}_t \;\; \mathbf{x} = \delta \tag{2.4}$$

for single input or single output systems, and

$$\mathbf{M}_t c(\mathbf{x}) = \delta \tag{2.5}$$

for multiinput and multioutput (MIMO) systems. The matrix $\mathbf{M}_t$ is completely determined by the plant parameters, which we consider fixed here, and by the order t of the controller. The reader is referred to equations (4.2.13) and (4.2.25) for the exact form of $\mathbf{M}_t$. Let $R(\mathbf{M}_t)$ denote the image of the map $\mathbf{M}_t$.

<u>Theorem 2.1</u>

<u>Let the plant be single input (multioutput) or single output (multiinput) and let $\mathbf{M}_t$ be defined as above. There exists a $\mathrm{t^{th}}$ order stabilizing controller if and only if</u>

$$\underline{R(\mathbf{M}_t) \cap H_n \neq \phi} \;. \tag{2.6}$$

For the MIMO case equation (2.5) shows that (2.6) is also a necessary condition for stabilizability with a tth order controller.

As t ranges over $0, 1, 2, \cdots$, the size of the matrix $\mathbf{M}_t$ and the numerical values of its entries change but the form of the equation remains unchanged. If t is high enough that $R(\mathbf{M}_t) = R^{n+1}$, (2.6) is trivially satisfied and (2.4) has a solution for every δ. This corresponds to the arbitrary pole assignment case and will occur for $t \geq q - 1$ in SISO systems, and generically for $t \geq \frac{q-m}{m}$ in single input m output systems and $t \geq \frac{q-r}{r}$ in single output r input systems.

For low values of t the problem of checking (2.6) cannot be completely solved without grappling with the nonlinear Hurwitz conditions. Fortunately however, some useful necessary conditions for (2.6) can be obtained using a theorem of linear programming and this is described below.

Let $\mathbf{y} > 0$ denote that every component y_i is strictly positive and let

$$P_n^+ := \{\delta \in R^{n+1} | \delta > 0\} \tag{2.7}$$

and

$$P_n^- := \{\delta \in R^{n+1} | \delta < 0\} \tag{2.8}$$

denote all polynomials of degree n with strictly positive and strictly negative coefficients, respectively. Clearly,

$$H_n \subset P_n^+ \cup P_n^-. \tag{2.9}$$

Lemma 2.2

If there exists a tth order stabilizing controller then there exists $\mathbf{x}$ so that

$$\mathbf{M}_t\mathbf{x} > 0 . \tag{2.10}$$

Proof

Note that there exists $\mathbf{x}$ so that $\mathbf{M}_t\mathbf{x} > 0$ if and only if there exists $\mathbf{x}$ such that $\mathbf{M}_t\mathbf{x} < 0$. Therefore if (2.10) fails $R(\mathbf{M}_t) \cap P_n^+ = \phi$ and $R(\mathbf{M}_t) \cap P_n^- = \phi$ and so $R(\mathbf{M}_t) \cap H_n = \phi$.

The condition (2.10) can be checked by the following theorem [59].

Theorem 2.3(Gordan's Theorem of the Alternative)

For each given matrix A,

Either

I. $A\mathbf{x} > 0$ has a solution $\mathbf{x}$

or

II. $A^T\mathbf{y} = 0, \mathbf{y} \geq 0$ has a solution $\mathbf{y}$

but never both.

(Here $\mathbf{y} \geq 0$ denotes that at least one component of $\mathbf{y}$ is positive, and no component is negative).

Geometrically we may interpret Gordan's theorem as follows. Either there exists a vector $\mathbf{x}$ which makes a strictly acute angle ($< \frac{\pi}{2}$) with all the row vectors of A, Figure 2.1a, or the origin can be expressed as a nontrivial, nonnegative linear combination of the rows of A, Figure 2.1b.

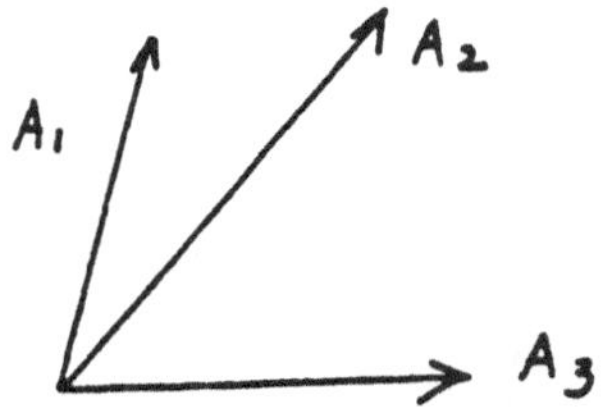

Figure 2.1a

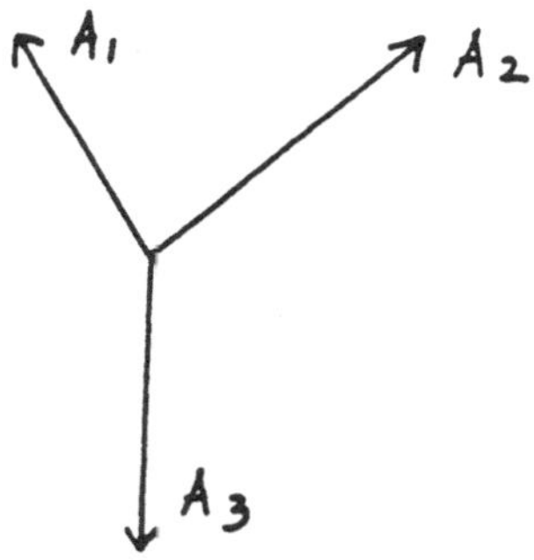

Figure 2.1b

Gordan's theorem leads to the following useful result on low order stabilization. The result is applicable to the general multiinput multioutput case.

Theorem 2.4

If

$$\mathbf{M}_t^T\mathbf{y} = 0, \qquad \mathbf{y} \geq 0$$

has a solution $\mathbf{y}$, then

$$R(\mathbf{M}_t) \cap H_n = \varphi$$

and no t^{th} order compensator can stabilize the given plant.

The condition given by Gordan's theorem can be checked by solving phase I (i.e. finding a feasible solution) of the linear programming problem:

$$\mathbf{M}_t^T\mathbf{y} = 0, \qquad \sum_j y_j = 1 \qquad \forall y_i \geq 0 \tag{2.11}$$

The condition $\mathbf{M}_t\mathbf{x} > 0$ can also be directly checked by a slight modification of the general linear programming problem, [60], which is defined as follows:

$$\text{Minimize (or Maximize)} f(x) = \sum_{j=1}^{n} c_j x_j \tag{2.12}$$

subject to

$$\sum_{j=1}^{n} a_{ij} x_j (\leq, =, \geq) b_i, \qquad i = 1, 2, \cdots, m,$$

$$x_j \geq 0, \qquad j = 1, 2, \cdots, n.$$

Therefore we modify (2.10) to

$$\underbrace{\left(\mathbf{M}_t \;\vdots\; -\mathbf{M}_t\right)}_{\mathbf{M}_n} \underbrace{\begin{pmatrix} \mathbf{y} \\ \cdots \\ \mathbf{z} \end{pmatrix}}_{\mathbf{x}_n} \geq \xi \tag{2.13}$$

and set up the problem as

$$\min f(x_n) = \sum_j x_{nj} \tag{2.14}$$

subject to

$$\sum_j m_{n(ij)} x_{nj} \geq \xi, \qquad i = 1, 2, \cdots\cdots,$$

$$x_{nj} \geq 0, \qquad j = 1, 2, \cdots\cdots$$

where $m_{n(ij)}$ denotes the $(i,j)^{th}$ element of $\mathbf{M}_n$. From the solution $\mathbf{x}_n$ of this problem, a solution $\mathbf{x}$ satisfying the inequality condition (2.10) is obtained with $\mathbf{x} = \mathbf{y} - \mathbf{z}$. It is possible to avoid the strict inequality in (2.10) by introducing a positive slack variable ξ. This slack variable may be chosen arbitrarily to be any positive vector without affecting the solvability of (2.10) as proved below.

Lemma 2.5

Let $\xi > 0$ be an arbitrary positive vector. Then

$$\mathbf{M}_t\mathbf{x} > 0 \qquad \text{has a solution} \tag{2.15}$$

if and only if

$$\mathbf{M}_t\mathbf{x} \geq \xi \qquad \text{has a solution}\ . \tag{2.16}$$

Proof

The proof is by contradiction. Let us suppose that $\xi > 0$ and $\xi^* > 0$ are fixed vectors such that

$$\mathbf{M}_t\mathbf{x} \geq \xi = \begin{pmatrix} \xi_1 \\ \xi_2 \\ \vdots \\ \xi_n \end{pmatrix} \qquad \text{has a solution} \tag{2.17}$$

but

$$\mathbf{M}_t\mathbf{y} \geq \xi^* = \begin{pmatrix} \xi_1^* \\ \xi_2^* \\ \vdots \\ \xi_n^* \end{pmatrix} \qquad \text{has no solution}\ . \tag{2.18}$$

Then let

$$\alpha = \frac{\max(\xi_1^*, \cdots, \xi_n^*)}{\min(\xi_1, \cdots, \xi_n)} \tag{2.19}$$

and consider the vector $\alpha\xi^*$. Clearly

$$\min(\alpha\xi_1, \cdots, \alpha\xi_n) = \max(\xi_1^*, \cdots, \xi_n^*). \tag{2.20}$$

Now $\alpha\xi \geq \xi^*$ so that

$$\mathbf{M}_t(\alpha\mathbf{x}) \geq \alpha\xi \geq \xi^* \tag{2.21}$$

which contradicts (2.18). $\diamond$

3.SUFFICIENT CONDITIONS USING THE STABILITY HYPERSPHERE

The results of the last section show how linear programming can be used to obtain necessary conditions for stabilizability by a tth order controller. If these necessary conditions are satisfied there exists $\mathbf{x}$ such that $\mathbf{M}_t\mathbf{x} \in P_n^+$. In low order problems it will frequently be true that this solution also satisfies $\mathbf{M}_t\mathbf{x} \in H_n$ or $\mathbf{M}_t c(\mathbf{x}) \in H_n$ and then $\mathbf{x}$ represents a stabilizing controller. In general, however, this will not happen, and this motivates us to develop a sufficient condition for (2.6) to hold. To state this result let $\delta \in R^{n+1}$ be Hurwitz and let $\rho(\delta)$ denote the Euclidean radius of the largest stability hypersphere centered at δ as in Theorem 1.3.2. Let $d(\delta, \mathbf{M}_t)$ denote the Euclidean distance between δ and the subspace $R(\mathbf{M}_t)$.

Theorem 3.1

Let δ denote any Hurwitz vector. If

$$\frac{d\,(\delta,\ \mathbf{M}_t)}{\rho(\delta)} < 1 \tag{3.1}$$

then the orthogonal projection δ_M of δ onto $R(\mathbf{M}_t)$ satisfies

$$\delta_M \in R(\mathbf{M}_t) \cap H_n. \tag{3.2}$$

Proof

The proof is obvious from the geometrical construction shown below in Figure 3.1. The condition (3.1) guarantees that the stability hypersphere centered at δ intersects the subspace $R(\mathbf{M}_t)$. $\diamond$

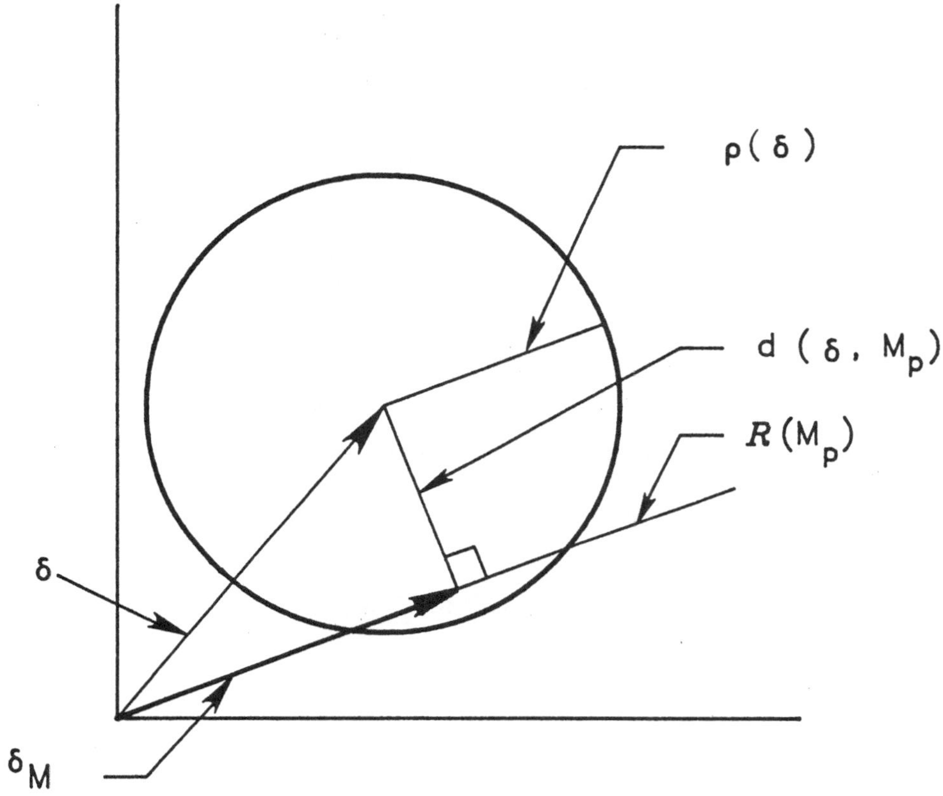

Illustration of Theorem 6.3.1

Figure 3.1

We assume that $\mathbf{M}_t$ has full rank. This entails no loss of generality as $\mathbf{x}$ can be redefined, if necessary to make it so. The equation for $d(\delta, \mathbf{M}_t)$ then, is

$$d(\delta,\ \mathbf{M}_t) = \|\mathbf{M}_t(\mathbf{M}_t^T\mathbf{M}_t)^{-1}\mathbf{M}_t^T\delta - \delta\|_2. \tag{3.3}$$

The above theorem suggests the following minimization problem:

$$\min_{\delta \in H_n} \frac{d(\delta,\ \mathbf{M}_t)}{\rho(\delta)} \tag{3.4}$$

Once we find a δ which satisfies (3.1) we can project δ orthogonally onto $R(\mathbf{M}_t)$ to obtain the vector δ_M which is both Hurwitz and in the subspace $R(\mathbf{M}_t)$. Then

$$\delta_M = \mathbf{M}_t(\mathbf{M}_t^T\mathbf{M}_t)^{-1}\mathbf{M}_t^T\delta \tag{3.5}$$

is the closed loop characteristic vector and

$$\mathbf{x} = (\mathbf{M}_t^T\mathbf{M}_t)^{-1}\mathbf{M}_t^T\delta \tag{3.6}$$

is the vector of transfer function coefficients of the stabilizing controller.

For the case

$$\mathbf{M}_t c(\mathbf{x}) = \delta \tag{3.7}$$

where $c(\mathbf{x})$ is a nonlinear function of the controller parameter $\mathbf{x}$ we may use the following strategy:

Write

$$\mathbf{M}_t\mathbf{y} = \delta \tag{3.8}$$

$$\mathbf{y} = c(\mathbf{x}) \tag{3.9}$$

Let δ be a nominal choice satisfying (3.1). Then let

$$\mathbf{y}^0 = (\mathbf{M}_t^T\, \mathbf{M}_t)^{-1}\mathbf{M}_t^T\delta := \mathbf{N}_t\delta \tag{3.10}$$

and let $\rho(\mathbf{y}^0)$ denote the radius of the largest stability hypersphere in the $\mathbf{y}$ space centered at $\mathbf{y}^0$. The calculation of $\rho(\mathbf{y}^0)$ was given in Theorem 2.3.1 of Chapter 2. Now if $\mathbf{x}$ is such that

$$\|\mathbf{y}^0 - c(\mathbf{x})\|_2 < \rho(\mathbf{y}^0) \tag{3.11}$$

it is clear that $\mathbf{M}_t c(\mathbf{x})$ is Hurwitz and a stabilizing controller has been found.

Theorem 3.2

Let δ be such that

$$\frac{d(\delta, \mathbf{M}_t)}{\rho(\delta)} < 1 \tag{3.12}$$

Then if $\mathbf{x}$ satisfies

$$\frac{\|\mathbf{N}_t\delta - c(\mathbf{x})\|_2}{\rho(\mathbf{N}_t\delta)} < 1 \tag{3.13}$$

it follows that

$$\mathbf{M}_t c(\mathbf{x}) \in H_n \tag{3.14}$$

and $\mathbf{x}$ corresponds to a tth order stabilizing controller.

The above theorem suggests the following algorithm:

$$\min_{\substack{\mathbf{x},\, \delta \\ \delta \in H_n}} \frac{\|\mathbf{N}_t\delta - c(\mathbf{x})\|_2}{\rho(\mathbf{N}_t\delta)} := J \tag{3.15}$$

The minimization of J can begin once a δ satisfying (3.12) has been found since otherwise $\rho(\mathbf{N}_t\delta)$ is not defined. If $J < 1$ is attained a stabilizing controller has been found.

4. EXAMPLES

Example 1

Let

$$G(s) = \begin{pmatrix} \dfrac{-2s^5 - 2s^4 - 20s^3 + s^2 + 14s - 14}{-s^5 - 10s^4 - 5s^3 - 6s^2 - 8s + 4} \\ \dfrac{2s^5 + s^4 - 20s^3 - 80s^2 - 80s - 14}{-s^5 - 10s^4 - 5s^3 - 6s^2 - 8s + 4} \end{pmatrix}$$

For a 0th order controller we have

$$\mathbf{M}_0 = \begin{pmatrix} -1 & -2 & 2 \\ -10 & -2 & 1 \\ -5 & -20 & -20 \\ -6 & 1 & -80 \\ -8 & 14 & -80 \\ 4 & -14 & -14 \end{pmatrix}.$$

Using linear programming we obtained $\mathbf{M}_0\mathbf{x} > 0$ with the controller parameter

$$\mathbf{x} = \begin{pmatrix} -0.001 \\ -1 \\ -1 \end{pmatrix}.$$

It turns out that $\mathbf{M}_0\mathbf{x}$ is Hurwitz. We verify that the roots of the closed loop system corresponding to this $\mathbf{x}$ are

$$\begin{pmatrix} -0.480749 \pm j0.674802 \\ -1.07459 \\ -39.1735 \\ -968.790 \end{pmatrix}$$

and therefore

$$C(s) = \begin{pmatrix} \frac{-1}{-0.001} & \frac{-1}{-0.001} \end{pmatrix}$$

is a stabilizing controller. A pole placement controller would be of second order, with the corresponding dimension of $\mathbf{x}$ being 9.

Example 2

Consider the following plant:

$$G(s) = \frac{-6.09s^{14} + 85.33s^{13} - 415.33s^{12} + 1094.9275s^{11}}{s^{15} - 22.08s^{14} + 29.47s^{13} - 71.95s^{12} - 5025s^{11}}$$

$$\frac{+11051.56s^{10} - 192s^{9} - 156.25s^{8}}{-11712.94s^{10} + 36.18s^{9} + 22.74s^{8}}$$

$$\frac{-419.68s^{7} + 1069.56s^{6} - 1298.42s^{5} + 375.1225s^{4}}{-191.64s^{7} - 907.8s^{6} + 1105.61s^{5}}$$

$$\frac{-1365.35s^{3} + 259.25s^{2} - 62.22s + 198.096}{+1322.45s^{3} - 259.7s^{2} - 3.34s - 208.55} .$$

For a 0th order controller

$$M_0 = \begin{pmatrix} 1 & 0 \\ -22.08 & -6.09 \\ 29.47 & 85.33 \\ -71.95 & -415.33 \\ -5025 & 1094.9275 \\ -11712.94 & 11051.56 \\ 36.18 & -192 \\ 22.74 & -156.25 \\ -191.64 & -419.68 \\ -907.8 & 1069.56 \\ 1105.61 & -1298.42 \\ 0 & 375.1225 \\ 1322.45 & -1365.35 \\ -259.7 & 259.25 \\ -3.34 & -62.22 \\ -208.55 & 198.096 \end{pmatrix} .$$

Using linear programming we have $\mathbf{M}_0^T\mathbf{y} = 0, \qquad \mathbf{y} \geq 0$ with

$$\mathbf{y} = \begin{pmatrix} 0 \\ 0 \\ 0 \\ 0 \\ 0 \\ 0.07915 \\ 0 \\ 0 \\ 0 \\ 0 \\ 0 \\ 0.21974 \\ 0.70109 \\ 0 \\ 0 \\ 0 \end{pmatrix}$$

as a solution to the conditions required by Gordan's Theorem. It follows from Gordan's Theorem and Theorem 2.4 that there is no 0th order stabilizing controller. We increase the order of the controller to 1. This gives

$$\mathbf{M}_1 = \begin{pmatrix} 1 & 0 & 0 & 0 \\ -22.08 & 1 & -6.09 & 0 \\ 29.47 & -22.08 & 85.33 & -6.09 \\ -71.95 & 29.47 & -415.33 & 85.33 \\ -5025 & -71.95 & 1094.9275 & -415.33 \\ -11712.94 & -5025 & 11051.56 & 1049.9275 \\ 36.18 & -11712.94 & -192 & 11051.56 \\ 22.74 & 36.18 & -156.25 & -192 \\ -191.64 & 22.74 & -419.68 & -156.25 \\ -907.8 & -191.64 & 1069.56 & -419.68 \\ 1105.61 & -907.8 & -1298.42 & 1069.56 \\ 0 & 1105.61 & 375.1225 & -1298.42 \\ 1322.45 & 0 & -1365.35 & 375.1225 \\ -259.7 & 1322.45 & 259.25 & -1365.35 \\ -3.34 & -259.7 & -62.22 & 259.25 \\ -208.55 & -3.34 & 198.096 & -62.22 \\ 0 & -208.55 & 0 & 198.096 \end{pmatrix}.$$

Using linear programming we found that $\mathbf{M}_1\mathbf{x} > 0$ for

$$\mathbf{x} = \begin{pmatrix} 0.001 \\ -0.030557 \\ -0.008807 \\ -0.031460 \end{pmatrix}.$$

However $\mathbf{M}_1\mathbf{x}$ is not Hurwitz. Therefore we adopt the minimization procedure (3.4). With an initial choice of $\delta^0 \in R^{17}$ corresponding to the polynomial $\delta^0(s) = (s+1)^{16}$ we find that $\rho(\delta^0) = 1$ and $d(\delta^0, \mathbf{M}_1) = 3568.24$, which does not satisfy the condition of Theorem 3.1. After "minimizing" (3.4) numerically we get a new value of δ which gives

$$d^*(\delta, \mathbf{M}_1) = 1.0 \times 10^{-8}$$

$$\rho^*(\delta) = 1.0 \times 10^{-2}.$$

Since $\frac{d^*(\rho,\mathbf{M}_1)}{\rho^*(\delta)} = 10^{-6} < 1$, we get from Theorem 3.1 the stabilizing controller parameter vector

$$\mathbf{x} = \begin{pmatrix} 0.00249987 \\ -0.0949889 \\ -0.0300101 \\ -0.9999985 \end{pmatrix}.$$

The roots of the closed loop system are

$$\begin{pmatrix} -0.0054836 \\ -0.2471693 \\ -0.2189355 \pm j0.7040592 \\ -0.5174567 \pm j0.8936120 \\ -1.0393600 \pm j0.2263056 \\ -0.9189383 \pm j0.6516824 \\ -0.2942842 \pm j1.3348327 \\ -0.5081362 \pm j1.6949029 \\ -2.8581280 \pm j2.8312120 \end{pmatrix}$$

and the corresponding 1st order stabilizing controller is

$$C(s) = \frac{-0.0300101s - 0.9999985}{0.00249987s - 0.0949889}.$$

A pole placement or observer based solution would be of 14th order with the corresponding dimension of the controller parameter vector $\mathbf{x}$ being 30.

Remarks

We have given some results and computational procedures that can aid the designer in generating low order solutions to the problem of feedback stabilization. Since no necessary and sufficient conditions for stabilizability with a fixed order controller are available as yet, these results are not final and it is our hope that they will stimulate further work on this problem.

It is clear that progress on this problem can result if a better understanding of the geometry of the Hurwitz region and efficient ways of dealing with the Hurwitz conditions can be developed. This would sharpen the algorithm given here.

CHAPTER 7

STATE SPACE DESIGN OF LOW ORDER REGULATORS

1. INTRODUCTION

In this chapter, we continue our treatment of the low order feedback stabilization problem by developing a state space based algorithm. This algorithm first attempts to stabilize the closed loop system with a fixed order controller. This corresponds to an extended output feedback stabilization problem. We attempt to solve this iteratively. At each iteration a state feedback matrix assigning a prescribed set of eigenvalues is found and this matrix is approximated by output feedback. This is done successively by readjusting the desired closed loop pole locations in the left half of the complex plane to minimize a performance index that measures the deviation of the actual eigenvalues from the desired ones. A low order solution is found by sequentially increasing the controller order until stabilization is achieved.

The algorithm that is given depends on the parametrization of the state feedback pole assignment problem derived in [44]. This is briefly described in the next section. In Section 3, the fixed order output feedback stabilization problem is formulated as an optimization problem and Section 4 describes how the performance index can be decreased by increasing the controller order. Examples are given in Section 5 and some of the gradient evaluations of Section 4 are derived in the Appendix.

2. THE SYLVESTER EQUATION FORMULATION

An algorithm was introduced in [44] for solving the pole assignment problem using state feedback. This algorithm consists of solving for X and then for F

$$AX - X\tilde{A} = -BG \tag{2.1}$$

$$FX = G \tag{2.2}$$

for given $(A, B, \tilde{A})$ with an arbitrary choice of G. In (2.1) and (2.2) A, X and $\tilde{A}$ are $n \times n$ matrices. From a result in [61] the solution X of (2.1) generically has full rank if (A, B) is controllable and $(G, \tilde{A})$ is observable. Let $\lambda_i(T)$ denote the ith eigenvalue of T and $\lambda(T)$ the spectrum or eigenvalue set of T. It follows that if X has full rank the solution F has the property:

$$\lambda(A + BF) = \lambda(\tilde{A}) . \tag{2.3}$$

The advantages of this algorithm are:

a) The algebraic variety $F(\Lambda)$ of matrices F which assign a prescribed set of eigenvalues Λ can be obtained by setting $\Lambda = \lambda(\tilde{A})$ for a fixed $\tilde{A}$, and letting the free parameter G run through the set of all possible real values.

b) Efficient numerical procedures [62] are available for the solution of Sylvester's equation (2.1).

Based on this parameterization of $F(\Lambda)$ algorithms were given in [19] and [45] for optimizing the conditioning of the closed loop eigenvectors and in [46] for minimizing the norm of the state feedback matrix F. Here, we extend these results by considering measurement rather than state feedback and by treating the problem of stabilization rather than arbitrary pole placement.

3. OUTPUT FEEDBACK CONTROLLERS

Consider the linear time invariant plant S cascaded with the pth order feedback compensator C.

$$\begin{aligned} S : \dot{x} &= Ax + Bu \\ y_m &= Cx. \end{aligned} \tag{3.1}$$

$$\begin{aligned} C : \dot{x}_c &= A_c x_c + B_c y_m \\ u &= C_c x_c + D_c y_m \end{aligned} \tag{3.2}$$

The closed loop system is

$$\begin{pmatrix} \dot{x} \\ \dot{x}_c \end{pmatrix} = \begin{pmatrix} A + BD_cC & BC_c \\ B_cC & A_c \end{pmatrix} \begin{pmatrix} x \\ x_c \end{pmatrix} \tag{3.3}$$

or

$$\underbrace{\begin{pmatrix} \dot{x} \\ x_c \end{pmatrix}}_{\dot{x}_p} = \left\{ \underbrace{\begin{pmatrix} A & 0 \\ 0 & 0_p \end{pmatrix}}_{A_p} + \underbrace{\begin{pmatrix} B & 0 \\ 0 & I_p \end{pmatrix}}_{B_p} \underbrace{\begin{pmatrix} D_c & C_c \\ B_c & A_c \end{pmatrix}}_{K_p} \underbrace{\begin{pmatrix} C & 0 \\ 0 & I_p \end{pmatrix}}_{C_p} \right\} \underbrace{\begin{pmatrix} x \\ x_c \end{pmatrix}}_{x_p} \tag{3.4}$$

and the transfer function of the pth order compensator is

$$C(s) := C_c(sI - A_c)^{-1}B_c + D_c \; . \tag{3.5}$$

The formula (3.4) shows that any fixed order compensator design problem is equivalent to a static output feedback problem. In particular the problem of stabilization with a fixed order controller p is equivalent to that of stabilizing $A_p + B_pK_pC_p$ by choice of K_p. The general solution of this problem is unknown. The best available results on the output feedack problem are those of Brasch and Pearson [57] and Kimura [58] which deal respectively with arbitrary eigenvalue assignment and "almost" arbitrary eigenvalue assignment.

Let Λ denote a symmetric set of $n+p$ complex numbers (i.e. complex numbers occur in complex conjugate pairs) and let

$$\underline{K}_p(\Lambda) := \{K_p | K_p \in R^{(m+p)x(r+p)}, \lambda(A_p + B_p K_p C_p) = \Lambda\} \tag{3.6}$$

where $A_p \in R^{(n+p)x(n+p)}, B_p \in R^{(n+p)x(m+p)}$, and $C_p \in R^{(r+p)x(n+p)}$ are as in (3.4).

The result of Brasch and Pearson [57] states that if (A, B, C) is controllable and observable with controllability index ν_c and observability index ν_o, and $p \geq \min\{\nu_c, \nu_o\}$ then $K_p(\Lambda) \neq \emptyset$ for every choice of Λ. The result of Kimura [58] states that if $p \geq n - m - r + 1$ then $\lambda(A_p + B_p K_p C_p)$ can be made arbitrarily close to any set Λ of $n+p$ symmetric complex numbers.

The upper bound on the order of a stabilizing controller established by the above results is in general too conservative. This stems from the fact that both results essentially require arbitrary pole placement. In fact for specific choices of the $n+p$ complex numbers Λ, $K_p(\Lambda)$ will "almost always" be empty unless p the compensator order is high. To lower the compensator order we therefore relax the specification of Λ in (3.6) to a simply connected region $\Omega \subset C^-$, and consider the family

$$\underline{K}_p(\Omega) = \{K_p | K_p \in R^{(m+p)x(r+p)}, \lambda(A_p + B_p K_p C_p) \subset \Omega \subset C^-\}. \tag{3.7}$$

It is reasonable to expect that $\underline{K}_p(\Omega)$ will in general be nonempty for values of p much less than the lower bounds given by the results of Brasch and Pearson or Kimura and numerical examples support this.

The effective characterization of the family $\underline{K}_p(\Omega)$ is an unsolved open problem. Our approach to this problem will be to consider the state feedback family

$$\underline{F}_p(\Omega) = \{F_p | F_p \in R^{(m+p)x(n+p)}, \lambda(A_p + B_p F_p) \subset \Omega \subset C^-\} \tag{3.8}$$

and determine an $F_p \in \underline{F}_p(\Omega)$ and then find K_p such that $\|F_p - K_pC_p\|$ is small in the hope that such a $K_p \in \underline{K}_p(\Omega)$. The advantage of this approach is that the family $\underline{F}_p(\Omega)$ can be characterized conveniently as shown later. For the remainder of this section we drop the subscript p for convenience.

In general, even if $\|F - KC\|$ is small it is not true that $\lambda(A + BF)$ and $\lambda(A + BKC)$ are close. The latter can be achieved by making the eigenstructure of $A + BF$ as orthonormal as possible. Let $\sigma_{max}(T)$ and $\sigma_{min}(T)$ denote the largest and smallest singular values of T. It is well known [62],[63] that the perturbation of the eigenvalues of the matrix $(A + BF)$ for changes in the entries is small if the condition number $k(X) := \|X\|_2\|X^{-1}\|_2 := \sigma_{\max}(X)/\sigma_{\min}(X)$ of the eigenvector matrix X is small. Let $F - KC := T$ so that $A + BKC = A + BF - BT$. Then using the formula in [63] we have

$$\begin{aligned} |\lambda_i(A + BKC) - \lambda_j(A + BF)| &\leq \|BT\|_2 k(X) \\ &\leq \|B\|_2\|T\|_2 k(X) \qquad (3.9) \\ &\leq \|B\|_2\|F - KC\|_F k(X) \end{aligned}$$

which shows that control over the eigenvalue locations of $A + BKC$ can be obtained only if both $\|F - KC\|$ and $k(X)$ are kept small. One way of doing this is to minimize

$$\begin{aligned} J &= \alpha_1 k(X) + \alpha_2\|F - KC\|_F^2 \\ &= \alpha_1 \frac{\sigma_{max}(X)}{\sigma_{min}(X)} + \alpha_2 \text{Trace}\{(F - KC)^T(F - KC)\} \qquad (3.10) \end{aligned}$$

by letting $\lambda(A + BF)$ range over the region $\Omega \subset C^-$. Similarly, by letting $A + F_DC = A + BK_DC$ a dual problem can be formulated as

$$J_D = \beta_1 \frac{\sigma_{max}(X_D)}{\sigma_{min}(X_D)} + \beta_2 \text{Trace}\{(F_D - BK_D)^T(F_D - BK_D)\}. \qquad (3.11)$$

The idea of simultaneously improving the conditioning of the eigenstructure and of minimizing the norm of $F - KC$ was first introduced in Keel and Bhattacharyya [64],[65]. Here an improved version of this algorithm is presented. In particular we convert the constrained optimization problem to an unconstrained problem and extend the class of regions $\Omega \subset C^-$ to more general and useful regions. These details are given next.

4. STABILIZATION ALGORITHM

In the Sylvester equation approach described in Section 2,

$$AX - X\tilde{A} = -BG \tag{4.1}$$

$$FX = G \tag{4.2}$$

and let $\lambda(\tilde{A}) \subset \Omega \subset C^-$. Under the assumption $\lambda(A) \cap \lambda(\tilde{A}) = \emptyset$ and (A, B) controllable, $(G, \tilde{A})$ observable, the unique solution X will 'almost surely' be non-singular by deSouza and Bhattacharyya [61] and then $\lambda(A + BF) = \lambda(\tilde{A})$ with $F = GX^{-1}$. By letting $\lambda(\tilde{A})$ range over Ω and G the free parameter run through <u>all</u> possible values this formula generates the family $\underline{F}(\Omega)$ defined in (3.8).

If $\tilde{A}$ is a complex diagonal matrix in (4.1), it is clear that X in (4.1) is the corresponding complex eigenvector matrix. However it is desirable that these matrices be real for computational convenience. The following Lemma 4.5 shows that $\tilde{A}$ can be taken as a real matrix without loss of generality. Before we state Lemma 4.3 it is necessary to introduce some facts.

<u>**Definition 4.1**</u>

A real square matrix A is called a pseudo diagonal matrix if it is of the form

$$A = \begin{pmatrix} \alpha_1 & \beta_1 & & & & \\ -\beta_1 & \alpha_1 & & & & \\ & & \alpha_2 & \beta_2 & & \\ & & -\beta_2 & \alpha_2 & & \\ & & & & \alpha_3 & \\ & & & & & \ddots \end{pmatrix} \tag{4.3}$$

with α_i, β_i real.

Definition 4.2

A complex square matrix is called normal if $A^*A = AA^*$.

Lemma 4.3 [54]

A complex square matrix is unitary similar to a diagonal complex matrix if and only if it is normal.

Lemma 4.4

Any real pseudo diagonal matrix is normal.

Proof

Taking the ith block from (4.3) such as

$$A_i = \begin{pmatrix} \alpha_i & \beta_i \\ -\beta_i & \alpha_i \end{pmatrix} \tag{4.4}$$

we have

$$A_iA_i^* = \begin{pmatrix} \alpha_i^2 + \beta_i^2 & 0 \\ 0 & \alpha_i^2 + \beta_i^2 \end{pmatrix} = A_i^*A_i. \tag{4.5}$$

Thus, each block is normal. Now let

$$A = \mathrm{Diag}\,(A_1 \quad A_2 \quad \cdots \quad \cdots \quad A_n) \tag{4.6}$$

$$AA^* = \mathrm{Diag}\,(\,A_1A_1^* \quad A_2A_2^* \quad \cdots \quad A_nA_n^*\,) \tag{4.7}$$

$$A^*A = \mathrm{Diag}\,(\,A_1^*A_1 \quad A_2^*A_2 \quad \cdots \quad A_n^*A_n\,)\;. \tag{4.8}$$

Since $AA^* = A^*A$, the statement is true. ◊

Lemma 4.5

Let $(A + BF)X = X\tilde{A}$ and $(A + BF)Y = Y\hat{A}$ where

1. A, B, $\tilde{A}$, X and F are real matrices with appropriate dimensions,
2. $\tilde{A}$ is real pseudo diagonal, $\hat{A}$ is complex diagonal, and
3. X and Y are nonsingular.

Then,

$$k(X) = k(Y)\;. \tag{4.9}$$

Proof

From Lemma 4.3 and 4.4, $\hat{A}$ is known to be normal and unitary similar to the complex diagonal matrix $\tilde{A}$. Thus

$$\tilde{A} = U\hat{A}U^*. \tag{4.10}$$

Write

$$(A + BF)X = X\tilde{A} = XU\hat{A}U^* \tag{4.11}$$

so that

$$(A + BF)XU = XU\tilde{A} \tag{4.12}$$

and

$$XU = Y. \tag{4.13}$$

Now,

$$k(X) = k(XU) = k(Y). \quad \diamond \tag{4.14}$$

From this Lemma, minimizing $\sigma_{max}(X)/\sigma_{min}(X)$ in (4.10) is equivalent to minimizing $\sigma_{max}(Y)/\sigma_{min}(Y)$. Therefore. we can henceforth take $\tilde{A}$ as a real pseudo diagonal matrix without loss of generality.

In order to use a gradient based descent algorithm the closed form expression of the gradient of the performance index (3.10) with respect to the variables G, K and the variable elements of $\tilde{A}$ denoted $\tilde{a}_i$ is evaluated. The derivation is given in the Appendix.

Theorem 4.6

Given the performance index J in (3.10), and constraints (4.1) and (4.2), the gradients of J with respect to the independent variables G, K, and $\tilde{A}$ are as follows:

(a)

$$\frac{\partial J}{\partial G} = 2\{\alpha_2(F - KC)X^{-T} + B^T U^T\} \tag{4.15}$$

where U satisfies

$$\tilde{A}U - UA = \frac{\alpha_1}{\sigma^2_{min}(X)}\{\sigma_{min}(X)v_a u_a^T - \sigma_{max}(X)v_i u_i^T\}$$

$$-2\alpha_2 X^{-1}(F^T - (KC)^T)F \tag{4.16}$$

where v_a and u_a are right and left singular vectors corresponding to $\sigma_{max}(X)$ and v_i and u_i correspond to $\sigma_{min}(x)$, respectively.

(b) Let $\tilde{a}_i$ denote a variable element of $\tilde{A}$:

$$\frac{\partial J}{\partial \tilde{a}_i} = -\text{Trace}\left\{UX\frac{\partial \tilde{A}}{\partial \tilde{a}_i}\right\} \tag{4.17}$$

where U satisfies (4.16)

(c)

$$\frac{\partial J}{\partial K} = -2\alpha_2(F - KC)C^T \tag{4.18}$$

Equations (4.15) – (4.18) are used to devise a gradient algorithm that iterates on the free parameters G, K and the entries of $\tilde{A}$ to reduce J. At each iteration of the algorithm we get $\tilde{A}_i, F_i$ and K_i. Since $\lambda(\tilde{A}_i) \subset \Omega$ we have $\lambda(A + BF_i) \subset \Omega$ for each i. However $\lambda(A + BK_iC)$ may or may not be in Ω for each i, and the algorithm is designed to update K_i to drive $\lambda(A + BK_iC)$ closer to $\lambda(\tilde{A}_i) = \lambda(A + BF_i)$ after some iterations.

The following structure of the closed loop eigenvalue matrix $\tilde{A}$ ensures stability without constraints during the iterations:

$$\tilde{A} = \begin{pmatrix} -\tilde{a}_1^2 & \tilde{a}_2 & & & & & \\ -\tilde{a}_2 & -\tilde{a}_1^2 & & & & & \\ & & -\tilde{a}_3^2 & \tilde{a}_4 & & & \\ & & -\tilde{a}_4 & -\tilde{a}_3^2 & & & \\ & & & & a_5 & & \\ & & & & & \ddots & \\ & & & & & & \ddots \end{pmatrix}.$$

Note that $\tilde{a}_i$ in the matrix $\tilde{A}$ are the only nonzero parameters and furthermore the stability requirement,$\lambda(\tilde{A}) \subset C^-$, is automatically satisfied without constraints, for <u>all</u> real values of $\tilde{a}_i$.

We can also parameterize $\tilde{A}$ in such a way that the desired closed loop eigenvalue locations are automatically confined to some useful region Ω as in Figures 4.1 and 4.2.

In choosing $\tilde{A}$, a maximal number of 2x2 blocks are included in the initial choice. As the algorithm evolves some of the off diagonal terms may become very small. At that point we start to vary the corresponding diagonal terms independently. In the damping ratio region described in Figure 4.2, θ is also a free parameter.

Marginal Stability Region

For this case we can simply modify the matrix $\tilde{A}$ to

$$\tilde{A} = \begin{pmatrix} -(\tilde{a}_1^2+\gamma) & \tilde{a}_2 & & & & & \\ -\tilde{a}_2 & -(\tilde{a}_1^2+\gamma) & & & & & \\ & & -(\tilde{a}_3^2+\gamma) & \tilde{a}_4 & & & \\ & & -\tilde{a}_4 & -(\tilde{a}_3^2+\gamma) & & & \\ & & & & \ddots & & \\ & & & & & \ddots & \\ & & & & & & \ddots \end{pmatrix}$$

with $\tilde{a}_i$ as the real variable parameters and γ is fixed. The eigenvalues of $\tilde{A}$ are all to the left of the line $Re(s) = -\gamma$.

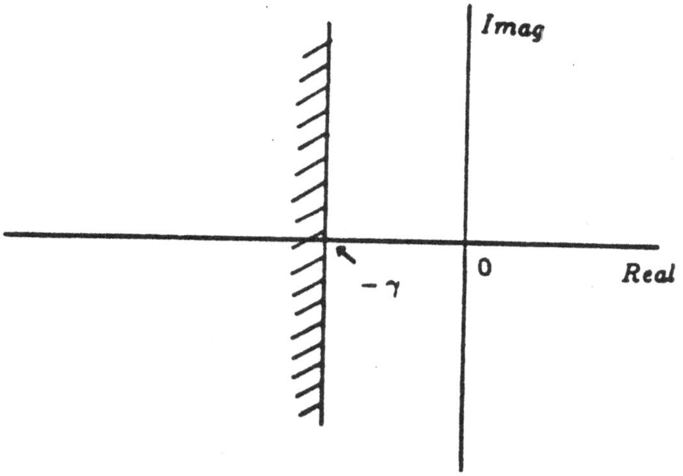

Figure 4.1

Marginal Stability Region

Damping Ratio Region

$$\tilde{A} = \begin{pmatrix} -(\bar{a}_1^2+\gamma) & (\bar{a}_1^2+\gamma)\tan\phi\sin\theta & & & \\ -(\bar{a}_1^2+\gamma)\tan\phi\sin\theta & -(\bar{a}_1^2+\gamma) & & & \\ & & \ddots & & \\ & & & \ddots & \\ & & & & \ddots \end{pmatrix}$$

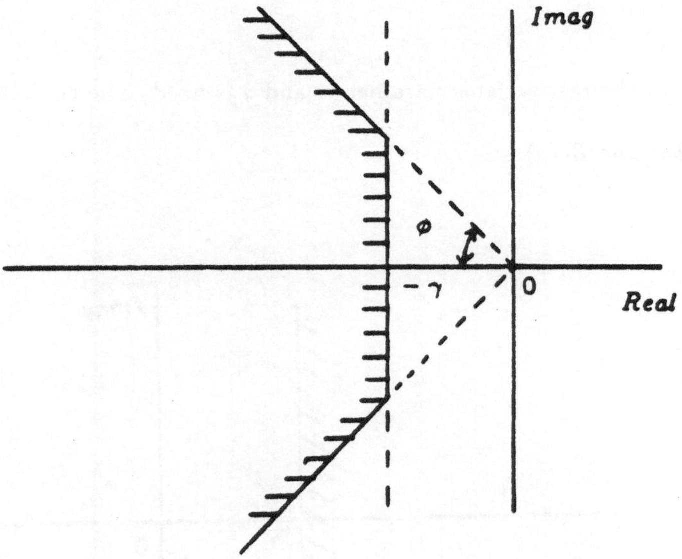

Figure 4.2

Damping Ratio Region

Now we discuss what happens when the proposed algorithm fails to find a stabilizing controller of order i. In this case, we increase the controller order to $i+1$. It is then necessary to have a way to select the initial values of G_0, $\bar{A}_0$ and K_0 for the controller of order $i+1$ to ensure that the performance index J keeps decreasing. The following theorem shows the way to select initial variables so that J always decreases with increasing controller order.

Theorem 4.7

Let J^* be the optimal performance index with optimal variables G^*, $\bar{A}^*$ and K^* where

$$J^* = \alpha_1, \frac{\sigma_{max}(X^*)}{\sigma_{min}(X^*)} + \alpha_2 \|F^* - K^* C\|_F^2 \tag{4.19}$$

and X^* and F^* satisfy

$$AX^* - X\bar{A}^* = -BG^*$$

$$F^* = G^*(X^*)^{-1} \ .$$

Then for the extended system

$$A_e = \begin{pmatrix} A & 0 \\ 0 & 0_i \end{pmatrix} \qquad B_e = \begin{pmatrix} B & 0 \\ 0 & I_i \end{pmatrix} \qquad C_e = \begin{pmatrix} C & 0 \\ 0 & I_i \end{pmatrix}, \tag{4.20}$$

the initial value of the corresponding performace index J_e is equal to J^* if the initial choice of the free variables are as follows:

$$G_e = \begin{pmatrix} G & 0 \\ 0 & X_3\bar{A}_i \end{pmatrix} \qquad K_e = \begin{pmatrix} K & 0 \\ 0 & X_3\bar{A}_iX_3^{-1} \end{pmatrix} \tag{4.21}$$

where $\bar{A}_i$ is an arbitrary pseudo diagonal submatrix of the extended matrix

$$\bar{A}_e = \begin{pmatrix} \bar{A} & 0 \\ 0 & \bar{A}_i \end{pmatrix}$$

and

$$X_3 = \begin{pmatrix} \sigma_1 & & & & \\ & \sigma_2 & & & \\ & & \ddots & & \\ & & & \ddots & \\ & & & & \sigma_i \end{pmatrix}$$

for arbitrary real numbers σ_i satisfying $\sigma_1 \geq \sigma_2 \geq \cdots \geq \sigma_i > 0$ with $\sigma_i \geq \sigma_{min}(X^*)$ and $\sigma_1 \leq \sigma_{max}(X^*)$.

Proof

Let the optimal values of J^* be obtained by G^* and K^*, then the extended system becomes

$$\underbrace{\begin{pmatrix} A & 0 \\ 0 & 0 \end{pmatrix}}_{A_e} \underbrace{\begin{pmatrix} X^* & X_1 \\ X_2 & X_3 \end{pmatrix}}_{X_e} - \underbrace{\begin{pmatrix} X^* & X_1 \\ X_2 & X_3 \end{pmatrix}}_{X_e} \underbrace{\begin{pmatrix} \bar{A}^* & 0 \\ 0 & \bar{A}_i \end{pmatrix}}_{\bar{A}_e}$$

$$= - \underbrace{\begin{pmatrix} B & 0 \\ 0 & I_i \end{pmatrix}}_{B_e} \underbrace{\begin{pmatrix} G^* & G_1 \\ G_2 & G_3 \end{pmatrix}}_{G_e}$$

$$= \begin{pmatrix} AX^* - X^*\bar{A}^* & AX_1 - X_1\bar{A}_i \\ -X_2\bar{A} & -X_3\bar{A}_i \end{pmatrix} = - \begin{pmatrix} BG & BG_1 \\ G_2 & G_3 \end{pmatrix}. \tag{4.22}$$

If we pick $G_1 = 0$ and $G_2 = 0$, then $X_1 = 0$ and $X_2 = 0$ and $X_3\bar{A}_i = G_3$. Here we choose

$$X_3 = \begin{pmatrix} \sigma_1 & & & \\ & \sigma_2 & & \\ & & \ddots & \\ & & & \sigma_i \end{pmatrix} \tag{4.23}$$

for arbitrary $\sigma_1 \geq \sigma_2 \geq \cdots \geq \sigma_i > 0$ with $\sigma_i \geq \sigma_{min}(X^*)$ and $\sigma_1 \leq \sigma_{max}(X^*)$. With this choice of X_3, we have

$$\sigma_{min}(X^*) = \sigma_{min}(X_e) \tag{4.24}$$

$$\sigma_{max}(X^*) = \sigma_{max}(X_e) \tag{4.25}$$

Therefore,

$$\frac{\sigma_{max}(X^*)}{\sigma_{min}(X^*)} = \frac{\sigma_{max}(X_e)}{\sigma_{min}(X_e)} \tag{4.26}$$

Now consider the term $\|F - KC\|_F^2$. Since

$$X_e = \begin{pmatrix} X & 0 \\ 0 & X_3 \end{pmatrix},$$

we have

$$X_e^{-1} = \begin{pmatrix} X^{-1} & 0 \\ 0 & X_3^{-1} \end{pmatrix} \tag{4.27}$$

where

$$X_3^{-1} = \begin{pmatrix} \frac{1}{\sigma_1} & & \\ & \ddots & \\ & & \frac{1}{\sigma_i} \end{pmatrix}.$$

Now

$$\begin{aligned} F_e = G_e X_e^{-1} &= \begin{pmatrix} G & 0 \\ 0 & X_3 \tilde{A}_i \end{pmatrix} \begin{pmatrix} X^{-1} & 0 \\ 0 & X_3^{-1} \end{pmatrix} \\ &= \begin{pmatrix} GX^{-1} & 0 \\ 0 & X_3 \tilde{A}_i X_3^{-1} \end{pmatrix}. \end{aligned} \tag{4.28}$$

Let

$$K_e = \begin{pmatrix} K^* & K_1 \\ K_2 & K_3 \end{pmatrix}. \tag{4.29}$$

Then

$$F_e - K_e C_e = \begin{pmatrix} GX^{-1} - KC & -K_1 \\ -K_2 C & X_3 \tilde{A}_i X_3^{-1} - K_3 \end{pmatrix}. \tag{4.30}$$

Here we choose $K_1 = 0$ and $K_2 = 0$. Also we can choose

$$K_3 = X_3 \tilde{A}_i X_3^{-1} \tag{4.31}$$

because X_3 and $\tilde{A}_i$ are well defined. With such a K we have

$$F_e - K_e C_e = \begin{pmatrix} GX^{-1} - KC & 0 \\ 0 & 0 \end{pmatrix} \tag{4.32}$$

Thus,

$$\|F - KC\|_F^2 = \|F_e - K_eC_e\|_F^2 \tag{4.33}$$

Therefore, we conclude that

$$\alpha_1 \frac{\sigma_{max}(X^*)}{\sigma_{min}(X^*)} + \alpha_2 \|F^* - K^*C\|_F^2 = \alpha_1 \frac{\sigma_{max}(X_e)}{\sigma_{min}(X_e)} + \alpha_2\|F_e - K_eC_e\|_F^2 \tag{4.34}$$

with the choices of

$$G_e = \begin{pmatrix} G^* & 0 \\ 0 & X_3\bar{A}_i \end{pmatrix} \quad \text{and} \quad K_e = \begin{pmatrix} K^* & 0 \\ 0 & X_3\bar{A}_iX_3^{-1} \end{pmatrix} \tag{4.35}$$

with X_3 as in (4.23). This concludes the proof. ◇

This theorem is useful for finding a low order stabilizing controller because it shows how by sequentially increasing the order of the controller, J can be guaranteed to decrease. Since a small enough value of each term of J confines the spectrum of $A + BKC$ to Ω (in accordance with (3.9))the algorithm eventually stabilizes the system.

5. EXAMPLES

The algorithm developed in the last section is applied to several examples here. The gradient calculations of Theorem 3.4 are used along with the Harwell subroutine package for optimization.

Example 1

The first example is a simplified model of the NASA F-8 Digital Fly-By-Wire(DFBW) airplane [66].Its dynamic equations are as follows:

$$\frac{d}{dt}\begin{pmatrix} p \\ r \\ \beta \\ \phi \end{pmatrix} = \begin{pmatrix} -2.6 & 0.25 & -38. & 0 \\ -0.075 & -0.27 & 4.4 & 0 \\ 0.078 & -0.99 & -0.23 & 0.052 \\ 1.0 & 0.078 & 0 & 0 \end{pmatrix}\begin{pmatrix} p \\ r \\ \beta \\ \phi \end{pmatrix} + \begin{pmatrix} 17. & 7. \\ 0.82 & -3.2 \\ 0 & 0.046 \\ 0 & 0 \end{pmatrix}\begin{pmatrix} \delta_a \\ \delta_r \end{pmatrix}$$

$$\begin{pmatrix} r \\ \phi \end{pmatrix} = \begin{pmatrix} 0 & 1 & 0 & 0 \\ 0 & 0 & 0 & 1 \end{pmatrix} \begin{pmatrix} p \\ r \\ \beta \\ \phi \end{pmatrix}$$

The given design specifications [66] are that the closed loop poles must be to the left of the line $s = -0.2$ i.e. $\gamma = 0.2$, and the damping factor is > 0.7,i.e. $\phi = \frac{\pi}{4}$ in Figure 4.2. For the optimization problem the initial values are chosen to be

$$\tilde{A}_0 = \begin{pmatrix} -3 & 2 & & \\ -2 & -3 & & \\ & & -5 & 3 \\ & & -3 & -5 \end{pmatrix}$$

$$G_0 = \begin{pmatrix} 1 & 1.5 & 0.5 & -2 \\ 5 & 1 & -0.25 & 0.5 \end{pmatrix}$$

$$K_0 = \begin{pmatrix} 0 & 0 \\ 0 & 0 \end{pmatrix}.$$

After 41 gradient iterations minimizing J in (3.10) the following 0th order stablilizing compensator is obtained.

$$K^* = \begin{pmatrix} 3.2451997 & -0.379821 \\ 2.8359286 & -0.055259 \end{pmatrix}$$

The corresponding data is shown in Tables 1.1, 1.2 and Figure 5.1. For comparison, the same problem was run without including the condition number term in J (i.e. $\alpha_1 = 0$ in (3.10)). It is seen from the corresponding data, shown in Table 1.3, 1.4 and Figure 5.2 that the condition number increases significantly, and although stabilization is achieved the closed loop eigenvalues fail to be in Ω.

TABLE 1.1

Eigenvalues for Example 1

$\alpha_1 = 1$, $\alpha_2 = 1$, $\phi = \frac{\pi}{4}$ $(\zeta \geq 0.7)$, $\gamma = 0.2$

$\lambda(A)$	$\lambda(A + BK_0C)$	$\lambda(A + BF^*)$	$\lambda(A + BK^*C)$
$-2.39 \pm j0.00$	$-2.39 \pm j0.00$	$-5.63 \pm j1.21$	$-2.78 \pm j2.49$
$+0.00 \pm j0.00$	$+0.00 \pm j0.00$	$-1.74 \pm j0.89$	$-3.01 \pm j0.00$
$-0.34 \pm j2.62$	$-0.34 \pm j2.62$		$-0.94 \pm j0.00$

TABLE 1.2

Performance Indices.

	$\|F - KC\|_F^2$	$k(X)$
Initial	61.3301	94.572
Optimal	19.8917	38.759

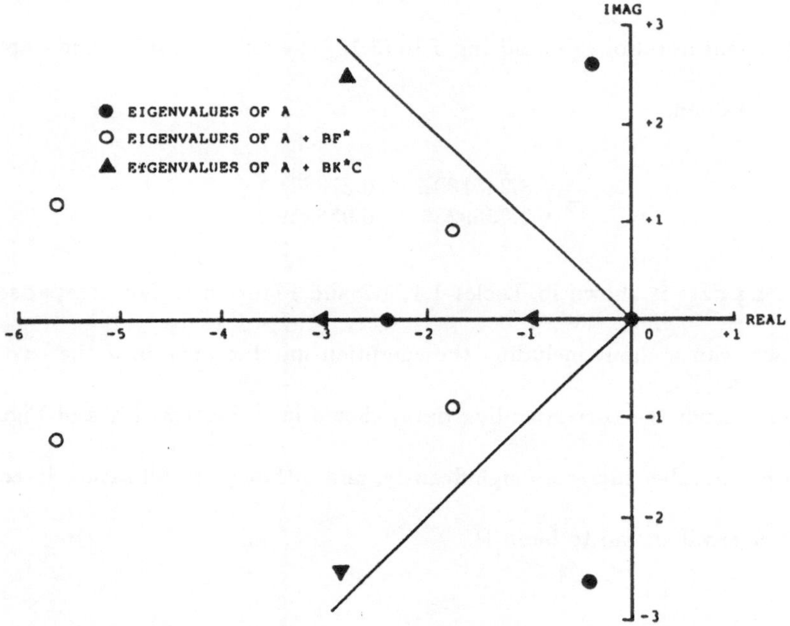

Figure 5.1

Eigenvalue locations corresponding to Table 1.1

TABLE 1.3

Eigenvalues for Example 1

($\alpha_1 = 0$, $\alpha_2 = 1$, $\phi = \frac{\pi}{4}$ ($\zeta \geq 0.7$), $\gamma = 0.2$)

$\lambda(A)$	$\lambda(A + BK_0C)$	$\lambda(A + BF^*)$	$\lambda(A + BK^*C)$
$-2.39 \pm j0.00$	$-2.39 \pm j0.00$	$-2.39 \pm j0.01$	$-1.44 \pm j2.54$
$+0.00 \pm j0.00$	$+0.00 \pm j0.00$	$-2.42 \pm j0.32$	$-3.44 \pm j0.00$
$-0.34 \pm j2.62$	$-0.34 \pm j2.62$		$-1.15 \pm j0.00$

TABLE 1.4

Performance Indices.

	$\|F - KC\|_F^2$	$k(X)$
Initial	61.3301	94.572
Optimal	0.01530	233089

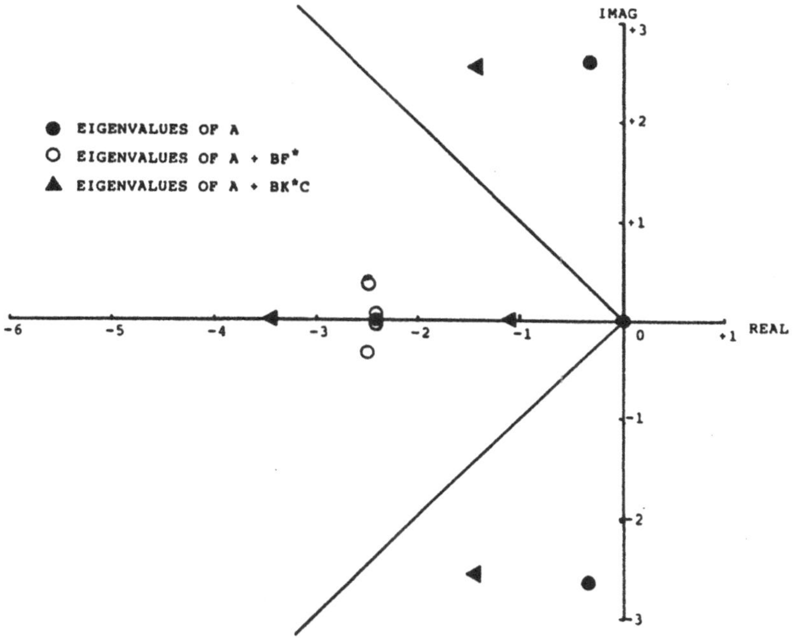

Figure 5.2

Eigenvalue locations corresponding to Table 1.3

Example 2

Consider the symmetric vibration model of the standard Draper/RPL satellite shown in Figure 5.3. The dynamic equations, taken from [67] are:

$$\frac{d}{dt}\begin{pmatrix} \theta \\ q_1 \\ q_2 \\ \dot{\theta} \\ \dot{q}_1 \\ \dot{q}_2 \end{pmatrix} = A \begin{pmatrix} \theta \\ q_1 \\ q_2 \\ \dot{\theta} \\ \dot{q}_1 \\ \dot{q}_2 \end{pmatrix} + B \begin{pmatrix} u_1 \\ u_2 \end{pmatrix}$$

$$\begin{pmatrix} y_1 \\ y_2 \end{pmatrix} = C \begin{pmatrix} \theta \\ q_1 \\ q_2 \\ \dot{\theta} \\ \dot{q}_1 \\ \dot{q}_2 \end{pmatrix}$$

where

$$A = \begin{pmatrix} 0 & 0 & 0 & 1 & 0 & 0 \\ 0 & 0 & 0 & 0 & 1 & 0 \\ 0 & 0 & 0 & 0 & 0 & 1 \\ 0 & 14.8732 & 32.8086 & 0 & 0 & 0 \\ 0 & -146.702 & -7476.64 & 0 & 0 & 0 \\ 0 & -41.8468 & -2699.36 & 0 & 0 & 0 \end{pmatrix}$$

$$B = \begin{pmatrix} 0 & 0 \\ 0 & 0 \\ 0 & 0 \\ -0.04168 & 0.23623 \\ 10.38611 & -25.647 \\ 3.725120 & -9.1629 \end{pmatrix}$$

$$C = \begin{pmatrix} 0 & 1 & 0 & 0 & 1 & 0 \\ 0 & 0 & 1 & 0 & 0 & 1 \end{pmatrix}$$

where

$q_1(q_2)$ is the vibration amplitude at $x = \frac{L}{2}(x = L)$.

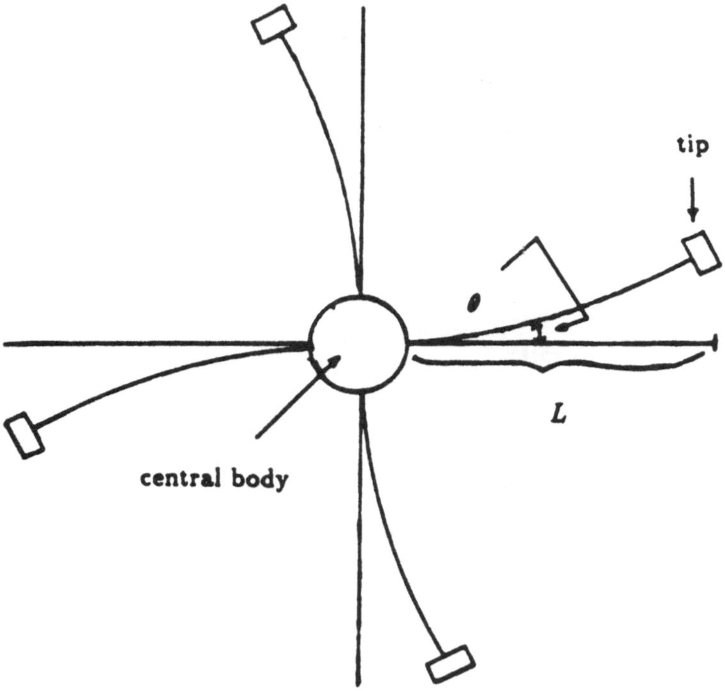

Draper/RPL symmetric vibrational model.

Figure 5.3

From the design specifications in [67], it follows that the closed loop system must have poles to the left of $s = -0.5$. For the minimization of J the initial values are chosen to be

$$\bar{A}_0 = \begin{pmatrix} -0.7 & 2 & & & & \\ -2 & -0.7 & & & & \\ & & -1 & 10 & & \\ & & -10 & -1 & & \\ & & & & -0.5 & 1 \\ & & & & -1 & -0.5 \end{pmatrix}$$

$$G_0 = \begin{pmatrix} 1.125 & 1.5 & -0.5 & 3.5 & 1.5 & 2 \\ -1 & 2.5 & 1.6 & 4 & 0.5 & -1 \end{pmatrix}$$

$$K_0 = \begin{pmatrix} 0 & 0 \\ 0 & 0 \end{pmatrix}.$$

After 67 iterations the following 0th order stabilizing controller is obtained:

$$K^* = \begin{pmatrix} -57.59521 & -482.41154 \\ -20.50957 & -195.78886 \end{pmatrix}.$$

Tables 2.1, 2.2 and Figure 5.4 display the performance indices and the corresponding eigenvalue locations. For the purpose of comparison, the problem was also run with the condition number term left out of the performance index (i.e. $\alpha_1 = 0$). In this case the algorithm failed to stabilize the system as shown in Tables 2.2, 2.4 and Figure 5.5. This example illustrates that both terms of the performance index need to be considered in the stabilization procedure.

TABLE 2.1

Eigenvalues for Example 2

$\alpha_1 = 1,\ \alpha_2 = 1,\ \gamma = 0.5$

$\lambda(A)$	$\lambda(A + BK_0C)$	$\lambda(A + BF^*)$	$\lambda(A + BK^*C)$
$+0.00 \pm j53.1$	$+0.00 \pm j53.1$	$-3.13 \pm j41.7$	$-0.86 \pm j46.0$
$+0.00 \pm j5.43$	$+0.00 \pm j5.43$	$-1.04 \pm j1.16$	$-0.94 \pm j5.48$
$+0.00 \pm j0.00$	$+0.00 \pm j0.00$	$-0.68 \pm j5.79$	$-1.02 \pm j1.16$
$+0.00 \pm j0.00$	$+0.00 \pm j0.00$		

TABLE 2.2

Performance Indices.

	$\|F - KC\|_F^2$	$k(X)$
Initial	11965506	409.2925
Optimal	60.20818	306.1105

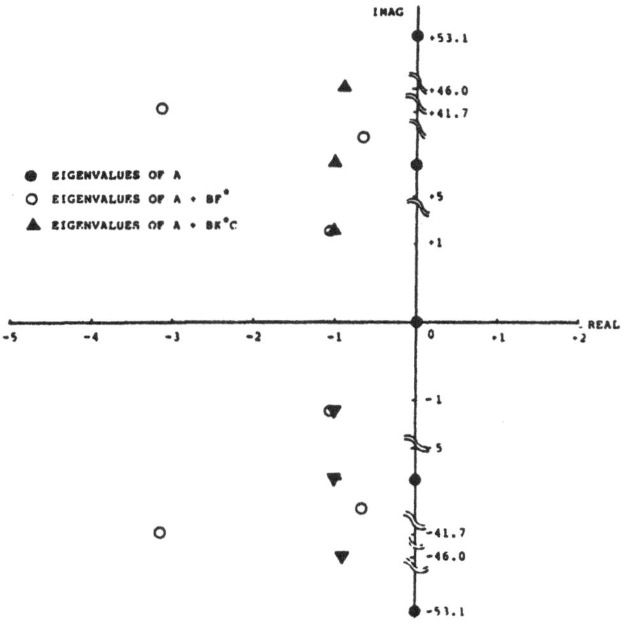

Figure 5.4

Eigenvalue locations corresponding to Table 2.1

TABLE 2.3

Eigenvalues for Example 2

$\alpha_1 = 0,\ \alpha_2 = 1,\ \gamma = 0.5$

$\lambda(A)$	$\lambda(A+BK_0C)$	$\lambda(A+BF^*)$	$\lambda(A+BK^*C)$
$+0.00 \pm j53.1$	$+0.00 \pm j53.1$	$-0.63 \pm j0.05$	$+178. \pm j0.00$
$+0.00 \pm j5.43$	$+0.00 \pm j5.43$	$-0.66 \pm j3.41$	$-2.57 \pm j6.19$
$+0.00 \pm j0.00$	$+0.00 \pm j0.00$	$-0.59 \pm j7.86$	$+1.61 \pm j0.00$
$+0.00 \pm j0.00$	$+0.00 \pm j0.00$		$+0.83 \pm j1.07$

TABLE 2.4

Performance Indices.

	$\|F-KC\|_F^2$	$k(X)$
Initial	11965506	409.2925
Optimal	490.5633	383369.1

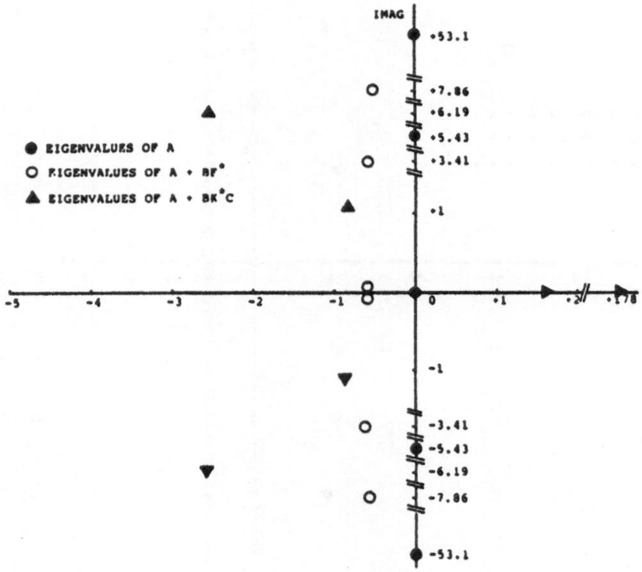

Figure 5.5

Eigenvalue locations corresponding to Table 2.3

APPENDIX

Proof of Theorem 3.4

(a)

$$J = \alpha_1 \frac{\sigma_{max}(X)}{\sigma_{min}(X)} + \alpha_2 \text{Trace}\{(F - KC)^T(F - KC)\}\,. \tag{A.1}$$

Let

$$\begin{aligned} J_1 :&= \frac{\sigma_{max}(X)}{\sigma_{min}(X)} \\ &= \text{Trace}\{\frac{\sigma_{max}(X)}{\sigma_{min}(X)}\} \end{aligned} \tag{A.2}$$

so that

$$\Delta J_1 = \text{Trace}\{\frac{1}{\sigma^2_{min}(X)}(\sigma_{min}(X)\Delta\sigma_{max}(X)) - \sigma_{max}\Delta\sigma_{min}(X)\}\,. \tag{A.3}$$

Note that

$$\Delta\sigma_{max}(X) = v_a u_a^T \Delta X \tag{A.4}$$

$$\Delta\sigma_{min}(X) = v_i u_i^T \Delta X \tag{A.5}$$

where v_i and u_i are left and right singular vectors corresponding to σ_{min} and v_a and u_a correspond to σ_{max}. Thus,

$$\Delta J_1 = \frac{1}{\sigma^2_{min}(X)} \text{Trace}\{\sigma_{min}(X) v_a u_a^T - \sigma_{max}(X) v_i u_i^T\}\Delta X\,. \tag{A.6}$$

Now

$$\begin{aligned} J_2 :&= \text{Trace}\{(F - KC)^T(F - KC)\} \\ &= \text{Trace}\{F^T F - (KC)^T F - F^T(KC) + (KC)^T(KC)\} \\ &= \text{Trace}(F^T F) - 2\text{Trace}\{(KC)^T F\} + \text{Trace}\{(KC)^T(KC)\} \end{aligned} \tag{A.7}$$

and

$$\Delta J_2 = 2\text{Trace}(F^T \Delta F) - 2\text{Trace}\{(KC)^T \Delta F\}$$
$$= 2\text{Trace}\{[F^T - (KC)^T]\Delta F\} \quad (A.8)$$

Now we have

$$\Delta J = \frac{\alpha_1}{\sigma_{min}^2(X)}$$
$$\text{Trace}\{\sigma_{min}(X) v_a u_a^T - \sigma_{max}(X) v_i u_i^T\}\Delta X \quad (A.9)$$
$$+ 2\alpha_2 \text{Trace}\{(F^T - (KC)^T)\Delta F\}.$$

From $F = GX^{-1}$, the gradient of F with respect to G is given directly as

$$\Delta F = \Delta G X^{-1} + G\Delta(X^{-1})$$
$$= \Delta G X^{-1} - G X^{-1} \Delta X X^{-1}$$
$$= \Delta G X^{-1} - F \Delta X X^{-1} \quad (A.10)$$
$$= (\Delta G - F\Delta X)X^{-1}.$$

Substituting $(A.10)$ into $(A.9)$ we have

$$\Delta J = 2\alpha_2 \text{Trace}\{(F^T - (KC)^T)\Delta G X^{-1}\}$$
$$+ \text{Trace}\{\frac{\alpha_1}{\sigma_{min}^2(X)}(\sigma_{min}(X) v_a u_a^T - \sigma_{max}(X) v_i u_i^T) \quad (A.11)$$
$$- 2\alpha_2 X^{-1}(F^T - (KC)^T)F\}\Delta X .$$

Since

$$A\Delta X - \Delta X \tilde{A} = -B\Delta G \quad (A.12)$$

we have

$$\Delta X = \sum_{i=1}^{n} \sum_{j=1}^{n} \gamma_{ij} A^{i-1} B \Delta G \tilde{A}^{j-1}. \quad (A.13)$$

Substituting $(A.13)$ into the second term of $(A.11)$ we have

$$\Delta J := 2\alpha_2 \mathrm{Trace}\{X^{-1}(F^T - (KC)^T)\Delta G\}$$

$$+ \mathrm{Trace}\{\sum_{i=1}^{n}\sum_{j=1}^{n}\gamma_{ij}\tilde{A}^{j-1}$$

$$\underbrace{(\frac{\alpha_1}{\sigma_{min}^2(X)}(\sigma_{min}(X)v_a u_a^T - \sigma_{max}(X)v_i u_i^T) - 2\alpha_2 X^{-1}(F^T - (KC)^T)F)}_{X_f}$$

$$A^{i-1}B\Delta G\} \qquad (A.14)$$

$$= 2\alpha_2 \mathrm{Trace}\{X^{-1}(F^T - (KC)^T)\Delta G\}$$

$$+ \mathrm{Trace}\{\underbrace{\sum_{i=1}^{n}\sum_{j=1}^{n}\gamma_{ij}\tilde{A}^{j-1}X_f A^{i-1}}_{U} B\Delta G\}$$

$$= \mathrm{Trace}\{2\alpha_2 X^{-1}(F^T - (KC)^T) + BU\}\Delta G\,.$$

From $(A.12)$ and $(A.13)$ it follows that U is the unique solution of

$$\tilde{A}U - UA = X_f. \qquad (A.15)$$

Therefore

$$\frac{\partial J}{\partial G} = 2\{\alpha_2(F - KC)X^{-T} + B^T U^T\} \qquad (A.16)$$

where U satisfies

$$\tilde{A}U - UA =$$

$$\frac{\alpha_1}{\sigma_{min}^2(X)}\{\sigma_{min}(X)v_a u_a^T - \sigma_{max}(X)v_i u_i^T\} - 2\alpha_2 X^{-1}(F^T - (KC)^T)F\,. \qquad (A.17)$$

(b)

Now we evaluate the gradients of (3.10) with respect to the variable elements of $\tilde{A}$. Recall the equation $(A.9)$

$$\Delta J = \frac{\alpha_1}{\sigma^2_{\min}(X)} \text{Trace}\{\sigma_{min}(X) v_a u_a^T - \sigma_{max}(X) v_i u_i^T\} \Delta X \\ + 2\alpha_2 \text{Trace}\{(F^T - (KC)^T)\Delta F\}. \tag{A.18}$$

From $F = GX^{-1}$, we compute(G is fixed)

$$\Delta F = -GX^{-1}\Delta X X^{-1} \\ = -F\Delta X X^{-1}. \tag{A.19}$$

Substituting ΔF into $(A.18)$

$$\Delta J = \text{Trace} \\ \underbrace{\{\frac{\alpha_1}{\sigma^2_{min}(X)}(\sigma_{min}(X) v_a u_a^T - \sigma_{max}(X) v_i u_i^T) - 2\alpha_2 X^{-1}(F - (KC)^T)F\}}_{X_f} \Delta X \tag{A.20}$$

$$= \text{Trace}\{X_f \Delta X\} \ .$$

Since

$$A\Delta X - \Delta X \tilde{A} = X \Delta \tilde{A} \tag{A.21}$$

$$\Delta X = \sum_{i=1}^{n} \sum_{j=1}^{n} \gamma_{ij} A^{i-1}(-X\Delta\tilde{A})\tilde{A}^{j-1}. \tag{A.22}$$

Substituting $(A.22)$ into $(A.20)$

$$\Delta J = \text{Trace}\{\sum_{i=1}^{n} \sum_{j=1}^{n} \gamma_{ij} X_f A^{i-1}(-X\Delta\tilde{A})\tilde{A}^{j-1}\} \\ = -\text{Trace}\{\underbrace{\sum_{i=1}^{n} \sum_{j=1}^{n} \gamma_{ij} \tilde{A}^{j-1} X_f A^{i-1}}_{U} X \Delta\tilde{A}\} \ . \tag{A.23}$$

It is clear that U is the unique solution of

$$\tilde{A}U - UA = X_f$$

as in $(A.14)$. Now,

$$\Delta J = -\text{Trace}\{UX\Delta\tilde{A}\}. \tag{A.24}$$

Therefore,

$$\frac{\partial J}{\partial \tilde{a}_i} = -\text{Trace}\{UX\frac{\partial \tilde{A}}{\partial \tilde{a}_i}\} \tag{A.25}$$

As an example the following calculation is considered. Let

$$U = \begin{pmatrix} u_{11} & u_{12} \\ u_{21} & u_{22} \end{pmatrix} \quad X = \begin{pmatrix} x_{11} & x_{12} \\ x_{21} & x_{22} \end{pmatrix} \quad \tilde{A} = \begin{pmatrix} \tilde{a}_1^2 & 0 \\ 0 & \tilde{a}_2^2 \end{pmatrix} \tag{A.26}$$

Then

$$\begin{aligned} \frac{\partial J}{\partial \tilde{a}_1} &= 2\text{Trace}\left\{\begin{pmatrix} u_{11} & u_{12} \\ u_{21} & u_{22} \end{pmatrix}\begin{pmatrix} x_{11} & x_{12} \\ x_{21} & x_{22} \end{pmatrix}\begin{pmatrix} 2\tilde{a}_1 & 0 \\ 0 & 0 \end{pmatrix}\right\} \\ &= 4\tilde{a}_1(u_{11}x_{11} + u_{12}x_{21}) \end{aligned} \tag{A.27}$$

$$\begin{aligned} \frac{\partial J}{\partial \tilde{a}_2} &= 2\text{Trace}\left\{\begin{pmatrix} u_{11} & u_{12} \\ u_{21} & u_{22} \end{pmatrix}\begin{pmatrix} x_{11} & x_{12} \\ x_{21} & x_{22} \end{pmatrix}\begin{pmatrix} 0 & 0 \\ 0 & 2\tilde{a}_2 \end{pmatrix}\right\} \\ &= 4\tilde{a}_2(u_{21}x_{12} + u_{22}x_{22}) \end{aligned} \tag{A.28}$$

or

$$\begin{pmatrix} \frac{\partial J}{\partial \tilde{a}_1} \\ \frac{\partial J}{\partial \tilde{a}_2} \end{pmatrix} = 4\begin{pmatrix} \tilde{a}_1(u_{11}x_{11} + u_{12}x_{21}) \\ \tilde{a}_2(u_{21}x_{12} + u_{22}x_{22}) \end{pmatrix} \tag{A.29}$$

(c)

Finally the gradient of J with respect to K is easily derived.

$$\begin{aligned} \Delta J &= -2\alpha_2\text{Trace}\{CF^T\Delta K - C(KC)^T\Delta K\} \\ &= -2\alpha_2\text{Trace}\{(CF^T - C(KC)^T)\Delta K\} \end{aligned} \tag{A.30}$$

Thus,

$$\frac{\partial J}{\partial K} = -2\alpha_2[F - KC]C^T. \tag{A.31}$$

CHAPTER 8

SUMMARY AND FUTURE RESEARCH

1. SUMMARY

This monograph has dealt with some problems related to the robust stability and robust stabilization of systems containing a real parameter vector subject to perturbation. Specifically, the following results have been given:

A. For systems where the closed loop characteristic polynomial coefficients are linear or affine in the parameter, we have

   i) calculated the largest stability hypersphere (Theorems 2.3.1 and 2.4.1) and the largest stability hyperellipsoid (Theorem 3.2.1) for the case of weighted perturbations, and

   ii) given constructive conditions for determining if a given perturbation region in parameter space is a stability region (Theorems 2.6.1, 3.3.1 and 3.3.2).

B. For the general case where the characteristic polynomial is a nonlinear function of the parameters, we have

   i) defined a stability margin (Section 3, Chapter 4)

   ii) given constructive sufficient conditions for determining if a given perturbation class is stabilized (Theorems 4.3.1 and 4.3.2, and Corollary 4.3.3), and

   iii) established a robustification procedure to design controllers that enlarge these stability regions (Section 4, Chapter 4).

C. For state space systems subject to structural perturbations, a stability hypersphere in parameter space has been determined using Lyapunov theory (Theorem 5.2.1) and a robustification algorithm based on this calculation has been developed (Section 2.3, Chapter5).

D. For the problem of stabilization with a low order controller, we have given a new lower bound on the order of a stabilizing controller using Gordan's theorem from linear programming (Theorem 6.2.4). Transfer function and state space based algorithms for low order stabilization have also been developed, respectively, in Chapters 6 and 7.

2. RESEARCH DIRECTIONS

The results described here are initial attempts and have many limitations. There exist many interesting open problems that need to be worked on. We single out some of them below:

1. The extension of Kharitonov's result (Theorem 1.3.1) to the parameter space.
2. The development of necessary and sufficient conditions for robust stability and stabilizability, in the general case, extending the results given here for the linear and affine cases.
3. The development of computational methods to check the robust stability conditions given, for instance, in Theorem 2.6.1, in geometric terms.
4. Effective ways of designing robust controllers, directly and nonconservatively, as opposed to the iterative methods given in Chapter 4 and 5. This is required because the solutions produced by the iterative methods are strongly dependent on the initial choice of the controller. The solution of this problem will require a much deeper understanding of the geometry of the Hurwitz region.
5. The direct inclusion of response specifications, as in the recent paper [68], into the

design of robust controllers.

6. The development of necessary and sufficient conditions for stabilizability with a fixed order controller, extending the results given here in Chapters 6 and 7. This could require an appropriate definition of the largest instability hypersphere in the controller parameter space, in a manner dual to the largest stability hypersphere defined, in Chapter 2, in the plant parameter space.

Some of these, in particular, items 1 and 2, are currently under study [69].

REFERENCES

[1] M.J. Chen and C.A. Desoer, "Necessary and Sufficient Conditions for Robust Stability of Linear Distributed Feedback Systems," *IEEE Trans. Automat. Contr.*, Vol. AC-29, pp. 880-888, Oct. 1984.

[2] H. Kimura, "Robust Stability of a Class of Transfer Functions," *IEEE Trans. Automat. Contr.*, pp. 788-793, Sept. 1984.

[3] N.A. Lehtomaki, N.R. Sandell and M. Athans, "Robustness Results in Linear Quadratic Gaussian Based Multivariable Control Designs," *IEEE Trans. Automatic Contr.*, Vol. AC-26, pp. 75-92, Feb. 1981.

[4] J.C. Doyle and G. Stein, "Multivariable Feedback Design: Concepts for a Classical/Modern Synthesis," *IEEE Trans. Automatic Contr.*, Vol. AC-26, pp. 4-16, Feb. 1981.

[5] V.L. Kharitonov, "Asymptotic Stability of an Equilibrium Position of a Family of Systems of Linear Differential Equations," *Differential. Uravnen.*, Vol. 14, No. 14, pp. 2086-2088, 1978.

[6] C.B. Soh, C.S. Berger, K.P. Dabke, "On the Stability Properties of Polynomials with Perturbed Coefficients," *IEEE Trans. Automat. Contr.*, Vol. AC-30, pp. 1033-1036, Oct. 1985.

[7] N.K. Bose, "A System-Theoretic Approach to Stability of Sets of Polynomials," *Contempory Mathematics*, Vol. 47, pp. 25-34, 1985.

[8] J.P. Guiver and N.K. Bose, "Strictly Hurwitz Property Invariance of Quartics Under Coefficient Perturbation," *IEEE Trans. Automat. Contr.*, Vol. AC-28, pp. 106-107,

1983.

[9] N.K. Bose, E.I. Jury and E. Zeheb, "On Robust Schur and Hurwitz Polynomials," *Proc. IEEE 1986 Conference on Decision and Contr.*, Dec. 1986, Athens, Greece.

[10] M.A. Argoun, "Allowable Coefficient Perturbations with Preserved Stability of a Hurwitz Polynomial," *Int. J. Control,* Vol. 44, No. 4, 927-934, 1986.

[11] S. Bialas and J. Garloff, "Stability of Polynomials Under Coefficient Perturbations," *IEEE Trans. Automat. Contr.*, Vol. AC-30, pp. 310-312, March. 1985.

[12] K.S. Yeung, "Linear System Stability Under Parameter Uncertainties," *Int. J. Control,* Vol. 38, No. 2, pp. 459-464, 1983.

[13] A.T. Fam and J. S. Meditch, "A Canonical Parameter Space for Linear Systems Design," *IEEE Trans. Automat. Contr.*, Vol. AC-23, No. 3, pp. 454-458, June 1978.

[14] A.T. Fam, "On the Geometry of Stable Polynomials of One and Two Variables," *Information Linkage Between Applied Mathematics and Industry,"* pp. 397-407, Academic Press, 1979.

[15] B.R. Barmish, "Invariance of the Strict Hurwitz Property for Polynomials with Perturbed Coefficients," *IEEE Trans. Automat. Contr.*, Vol. AC-29, pp. 935-936, Oct. 1984.

[16] R.M. Biernacki, "Sensitivities of Stability Constraints and Their Applications," *IEEE Trans. Automat. Contr.*, Vol. AC-31, pp. 642-689, July 1986.

[17] H. Hwang, M.S. Thesis, Dept. of Elec. Eng., "Robust Stabilization of Plants Subject to Structured Real Parameter Perturbations," Texas A&M University, July 1986.

[18] L.H. Keel, J.W. Howze and S.P. Bhattacharyya, "Robust Compensation Via Pole

Placement," *Proc. 1985 American Control Conference*, Boston, June 1985.

[19] L.H. Keel and S.P. Bhattacharyya, "Low Order Robust Stabilizer Design Using Hurwitz Conditions," *Proc. IEEE Conf. Decision and Control*, Dec. 1985, Ft. Lauderdale, Florida.

[20] R.M. Biernacki, H.S. Hwang and S.P. Bhattacharyya, "Robust Stabilization of Plants Subject to Real Parameter Perturbations," Technical Report No. 86, Dept. of Elec. Eng., Texas A&M University, June 1986.

[21] A. C. Bartlett, C. V. Hollot and Huang Lin, "Root Locations of an Entire Polytope of Polynomials: It Suffices to Check the Edges," Technical Report No. UMASS CCS 81-103, University of Massachussetts, Amherst, MA., 1986

[22] L. H. Keel, "Computer Aided Control System Design for Linear Timeinvariant Systems" Ph.D. Thesis, Dept. of Elec. Eng., Texas A&M University, College Station, Tx 77843, July 1986.

[23] L.H. Keel, S.P. Bhattacharyya and J.W. Howze, "Robust Control with Structured Perturbations," Technical Report #8603, Dept. of Elec. Eng., Texas A&M University, College Station, Tx, August 1986.

[24] R.K. Yedavalli, "Improved Measures of Stability Robustness for Linear State Space Models," *IEEE Trans. Aut. Control*, Vol. 30, No. 6, pp. 577-579, June 1985

[25] R.K. Yedavalli, "Perturbation Bounds for Robust Stability in Linear State Space Models," *Int. J. Control*, Vol. 42, pp. 1507-1517, Dec. 1985.

[26] R.K. Yedavalli, "Reduced Conservatism in Testing for Hurwitz Invariance of State Space Models," *IEEE Conf. Decision and Control*, Dec. 1985, Ft. Lauderdale,

Florida.

[27] R.K. Yedavalli, "Dynamic Compensator Design for Robust Stability of Linear Uncertain Systems," *Proc. IEEE Conf. on Decision and Control*, Athens Greece, Dec. 1986.

[28] R.V. Patel and M.Toda, "Quantitative Measures of Robustness for Multivariable Systems," *Proc. American Control Conference*, San Francisco, Ca. June 1985.

[29] B.R. Barmish, "Necessary and Sufficient Conditions for Quadratic Stabilizability of Uncertain Linear Systems," *Journal of Optimization Theory and Applications*, Vol. 46, 1985.

[30] B.R. Barmish and G. Leitmann, "On Ultimate Boundedness Control of Uncertain Systems in the Absence of Matching Conditions," *IEEE Trans. Automat. Contr.*, Vol. AC-27, pp. 253-258, Jan. 1982.

[31] J.S. Thorp and B.R. Barmish, "On Guaranteed Stability of Uncertain Linear Systems Via Linear Control," *Journal of Optimization Theory and Applications*, Vol. 35, pp. 559-579, 1981.

[32] B.R. Barmish, "Stabilization of Uncertain Systems Via Linear Control," *IEEE Trans. Automat. Contr.*, Vol. AC-28, pp. 848-850, 1983.

[33] D.D.Siljak *Nonlinear Systems: Parameter Analysis and Design*, Wiley, New York, 1969.

[34] J.S.Karmarkar and D.D.Siljak "A Computer Aided Design of Robust Regulators", pp.49-58, *Control Applications of Nonlinear Progamming*, Pergamon Press, 1979.

[35] C.V. Hollot, "Matrix Uncertainity Structures for Robust Stability," *Proc. American*

Control Conference, Boston, 1985.

[36] M.A. Argoun, "On Sufficient Conditions for the Stability of Interval Matrices," *Int. J. Control,* Vol. 44. No. 5, 1245-1250, 1986.

[37] John Doyle, "Analysis of Feedback Systems with Structured Uncertainties," *Proc. IEEE. Part D*, No. 6, Nov. 1982, pp. 242-250.

[38] R.R.E. de Gaston and M.G. Safonov, "Calculation of the Multiloop Stability Margin," *Proc. American Control Conference,* Seattle, Washington, June 1986.

[39] R.R.E. de Gaston and A. Sideris, "Multivariable Stability Margin Calculation with Uncertain Correlated Parameters," *Proc. IEEE Conf. on Decision and Control,* pp. 766-771, Dec. 1986. Greece.

[40] M. Saeki, "A Method of Robust Stability Analysis with Highly Structured Uncertainties," *IEEE Trans. Automat. Contr.*, Vol. AC-31, No. 10, pp. 925-940, Oct. 1986.

[41] J.D. Ackermann, "Design of Robust Controllers by Multi Model Methods," *Proc. of the 7th Int. MTNS Conference,* Stockholm, Sweden, June 1985.

[42] M. Vidyasagar, *Control System Synthesis: A Factorization Approach,* Cambridge, MA., MIT Press, 1985.

[43] S.P. Bhattacharyya, A.C. del Nero Gomes and J.W. Howze, "The Structure of Robust Disturbance Rejection Control," *IEEE Trans. Aut. Control*, Vol. 28, No. 5, pp. 874-881, Sept. 1983.

[44] S. P. Bhattacharyya and E. deSouza, "Pole Assignment via Sylvester's Equation," *Systems and Controls Letters*, Vol. 1, No. 4, pp. 261-263, Jan. 1982.

[45] R. K. Cavin III and S.P. Bhattacharyya, "Robust and Well Conditioned Eigenstructure Assignment via Sylvester's Equation," *Optimal Control: Applic. and Methods*, Vol. 4, pp. 205-212, 1983.

[46] L.H. Keel, J.A. Fleming and S.P. Bhattacharyya, "Minimum Norm Pole Assignment via Sylvester's Equation," *American, Mathematical Society Contemporary Mathematics Series*, Vol. 47, pp. 265-272, 1985.

[47] S.P. Bhattacharyya, L.H. Keel and J.W. Howze, "Feedback Stabilization with Low Order Regulators," Technical Report #8602, Dept. of Elec. Eng., Texas A&M University, College Station, TX. 77843, August 1986.

[48] M.J. Hopper, "Harwell Subroutine Library" Computer Science and Systems Division, AERE Harwell, Oxfordshire, Nov. 1981.

[49] S.P. Bhattacharyya and J.B. Pearson, "On Error Systems and the Servomechanism Problem," *Int. J. Control*, Vol. 15, No. 6, pp. 1041-1062, 1972.

[50] S.P. Bhattacharyya, "Disturbance Rejection in Linear Systems," *Int. J. System Science*, Vol. 5, No. 7, pp. 633-637, 1974.

[51] D. Ohm, J.W. Howze and S.P. Bhattacharyya, "Structural Synthesis of Multivariable Controllers,"*Automatica*, Vol. 21, No. 1, pp. 35-55, Jan. 1985.

[52] G.J. Franklin, J.D. Powell, and A. Emami-Naeini, *Feedback Control of Dynamic Systems*, Addison Wesley, Reading, Mass. 1986.

[53] S.N. Singh and A.R. Coelho, "Nonlinear Control of Mismatched Uncertain Linear Systems and Application to Control of Aircraft," *Journal of Dynamic Systems, Measurement and Control*, Vol. 106, pp. 203-210, Sept. 1984.

[54] P. Lancaster, *Theory of Matrices*, N.Y.: Academic Press, 1969.

[55] H. Kwakernaak and R. Sivan, *Linear Optimal Control Systems*, New York, N.Y.: Wiley-Interscience, 1972.

[56] D. G. Luenberger, "An Introduction to Observers," *IEEE Trans. Automat. Contr.*, Vol. AC-16, Dec. 1971.

[57] F.M. Brasch and J.B. Pearson, "Pole Placement Using Dynamic Compensator," *IEEE Trans. Automat. Contr.* Vol. AC-15, pp. 34-43, Feb. 1970.

[58] H. Kimura, "Pole Assignment by Gain Output Feedback," *IEEE Trans. Automat. Contr.*, Vol. AC-20, pp. 509-516, Aug. 1975.

[59] P. Gordan, "Über die Auflösungen linearer Gleichungen mit reelen coefficienten," *Mathematische Annalen,* Vol. 6, pp. 23-28, 1873.

[60] H.A. Taha, *Operational Reserach*, New York: MacMillan Publishing Co., 1971.

[61] E. deSouza and S.P. Bhattacharyya, "Controllability, Observability and the Solution of AX - XB =C," *Linear Algebra and Its Appli.*, Vol. 39, pp. 167-188, 1981.

[62] G.H. Golub and C.F. Van Loan, *Matrix Computations,* Baltimore, Maryland: The Johns Hopkins University Press, 1983.

[63] J.H. Wilkinson, *The Algebraic Eigenvalue Problem*, London: Oxford University Press, 1965.

[64] L.H. Keel and S.P. Bhattacharyya, "An Algorithm for Low Order Stabilizing Compesator Design via Sylvester's Equation," in *Proc. of the 2nd IEEE Contr. Sys. Soc. Symp. on CACSD*, Santa Barbara, CA, Mar. 1985.

[65] L.H. Keel and S.P. Bhattacharyya, "Compensator Design for Robust Eigenstructure

Assignment," in *Proc. of American Control Conference*, Boston, MA, June 1985.

[66] J.R. Elliott, "NASA's Advanced Control Law Program for the F-8 Digital-Fly-By-Wire Aircraft," *IEEE Trans. Automat. Contr.*, Vol. AC-22, pp. 753-757, Oct. 1977.

[67] D.S. Bodden and J.L. Junkins, "Eigenvalue Optimization Algorithms for Structural/ Controller Design Iterations," in *The Proc. of American Control Conference*, San Diego, CA, June 1984.

[68] Munther A. Dahleh and J.Boyd Pearson,Jr., "l^1-Optimal Feedback Controllers for MIMO Discrete-Time Systems" *IEEE Trans. Automat. Contr.*, Vol. AC-32, No.4, pp.314-322, April, 1987.

[69] Hervé Chapellat and S.P.Bhattacharyya, "Geometric Conditions for Robust Stability," Technical Report, Dept. of Elec. Eng., Texas A&M University, College Station, Texas 77843.

[70] F.R. Gantmacher, *Applications of the Theory of Matrices* Interscience Publishers Inc., New York, 1959.

Lecture Notes in Control and Information Sciences

Edited by M. Thoma and A. Wyner

Vol. 62: Analysis and Optimization of Systems
Proceedings of the Sixth International Conference on Analysis and Optimization of Systems
Nice, June 19–22, 1984
Edited by A. Bensoussan, J. L. Lions
XIX, 591 pages. 1984.

Vol. 63: Analysis and Optimization of Systems
Proceedings of the Sixth International Conference on Analysis and Optimization of Systems
Nice, June 19–22, 1984
Edited by A. Bensoussan, J. L. Lions
XIX, 700 pages. 1984.

Vol. 64: Arunabha Bagchi
Stackelberg Differential Games in Economic Models
VIII, 203 pages, 1984

Vol. 65: Yaakov Yavin
Numerical Studies in Nonlinear Filtering
VIII, 273 pages, 1985.

Vol. 66: Systems and Optimization
Proceedings of the Twente Workshop
Enschede, The Netherlands, April 16–18, 1984
Edited by A. Bagchi, H. Th. Jongen
X, 206 pages, 1985.

Vol. 67: Real Time Control of Large Scale Systems
Proceedings of the First European Workshop
University of Patras, Greece, Juli 9–12, 1984
Edited by G. Schmidt, M. Singh, A. Titli, S. Tzafestas
XI, 650 pages, 1985.

Vol. 68: T. Kaczorek
Two-Dimensional Linear Systems
IX, 397 pages, 1985.

Vol. 69: Stochastic Differential Systems – Filtering and Control
Proceedings of the IFIP-WG 7/1 Working Conference
Marseille-Luminy, France, March 12-17, 1984
Edited by M. Metivier, E. Pardoux
X, 310 pages, 1985.

Vol. 70: Uncertainty and Control
Proceedings of a DFVLR International Colloquium
Bonn, Germany, March, 1985
Edited by J. Ackermann
IV, 236 pages, 1985.

Vol. 71: N. Baba
New Topics in Learning Automata Theory and Applications
VII, 231 pages, 1985.

Vol. 72: A. Isidori
Nonlinear Control Systems: An Introduction
VI, 297 pages, 1985.

Vol. 73: J. Zarzycki
Nonlinear Prediction Ladder-Filters for Higher-Order Stochastic Sequences
V, 132 pages, 1985.

Vol. 74: K. Ichikawa
Control System Design based on Exact Model Matching Techniques
VII, 129 pages, 1985.

Vol. 75: Distributed Parameter Systems
Proceedings of the 2nd International Conference, Vorau, Austria 1984
Edited by F. Kappel, K. Kunisch, W. Schappacher
VIII, 460 pages, 1985.

Vol. 76: Stochastic Programming
Edited by F. Archetti, G. Di Pillo, M. Lucertini
V, 285 pages, 1986.

Vol. 77: Detection of Abrupt Changes in Signals and Dynamical Systems
Edited by M. Basseville, A. Benveniste
X, 373 pages, 1986.

Vol. 78: Stochastic Differential Systems
Proceedings of the 3rd Bad Honnef Conference, June 3–7, 1985
Edited by N. Christopeit, K. Helmes, M. Kohlmann
V, 372 pages, 1986.

Vol. 79: Signal Processing for Control
Edited by K. Godfrey, P. Jones
XVIII, 413 pages, 1986.

Vol. 80: Artificial Intelligence and Man-Machine Systems
Edited by H. Winter
IV, 211 pages, 1986.

Lecture Notes in Control and Information Sciences

Edited by M. Thoma and A. Wyner

Vol. 81: Stochastic Optimization
Proceedings of the International Conference, Kiew, 1984
Edited by I. Arkin, A. Shiraev, R. Wets
X, 754 pages, 1986.

Vol. 82: Analysis and Algorithms of Optimization Problems
Edited by K. Malanowski, K. Mizukami
VIII, 240 pages, 1986.

Vol. 83: Analysis and Optimization of Systems
Proceedings of the Seventh International Conference of Analysis and Optimization of Systems
Antiba, June 26-27, 1986
Edited by A. Bensoussan, J. L. Lions
XVI, 901 pages, 1986.

Vol. 84: System Modelling and Optimization
Proceedings of the 12th IFIP Conference
Budapest, Hungary, September 2–6, 1985
Edited by A. Prékopa, J. Szelezsán, B. Strazicky
XII, 1046 pages, 1986.

Vol. 85: Stochastic Processes in Underwater Acoustics
Edited by Charles R. Baker
V, 205 pages, 1986.

Vol. 86: Time Series and Linear Systems
Edited by Sergio Bittanti
XVII, 243 pages, 1986.

Vol. 87: Recent Advances in System Modelling and Optimization
Proceedings of the IFIP-WG 7/1 Working Conference
Santiago, Chile, August 27-31, 1984
Edited by L. Contesse, R. Correa, A. Weintraub
IV, 199 pages, 1987.

Vol. 88: Bruce A. Francis
A Course in H_∞ Control Theory
XI, 156 pages, 1987.

Vol. 89: G. K. H. Pang/A. G. J. McFarlane
An Expert System Approach to Computer-Aided Design of Multivariable Systems
XII, 223 pages, 1987.

Vol. 90: Singular Perturbations and Asymptotic Analysis in Control Systems
Edited by P. Kokotovic, A. Bensoussan, G. Blankenship
VI, 419 pages, 1987.

Vol. 91 Stochastic Modelling and Filtering
Proceedings of the IFIP-WG 7/1 Working Conference
Rome, Italy, Decembre 10-14, 1984
Edited by A. Germani
IV, 209 pages, 1987.

Vol. 92: L. T. Grujić, A. A. Martynyuk, M. Ribbens-Pavella
Large-Scale Systems Stability Under Structural and Singular Perturbations
XV, 366 pages, 1987.

Vol. 93: K. Malanowski
Stability of Solutions to Convex Problems of Optimization
IX, 137 pages, 1987.

Vol. 94: H. Krishna
Computational Complexity of Bilinear Forms
Algebraic Coding Theory and Applications to Digital Communication Systems
XVIII, 166 pages, 1987.

Vol. 95: Optimal Control
Proceedings of the Conference on Optimal Control and Variational Calculus
Oberwolfach, West-Germany, June 15-21, 1986
Edited by R. Bulirsch, A. Miele, J. Stoer and K. H. Well
XII, 321 pages, 1987.

Vol. 96: H. J. Engelbert/W. Schmidt
Stochastic Differential Systems
Proceedings of the IFIP-WG 7/1 Working Conference
Eisenach, GDR, April 6-13, 1986
XII, 381 pages, 1987.

Vol. 97: I. Lasiecka/R. Triggiani (Eds.)
Control Problems for Systems Described by Partial Differential Equations and Applications
Proceedings of the IFIP-WG 7.2 Working Conference
Gainesville, Florida, February 3-6, 1986
VIII, 400 pages, 1987.

Lecture Notes in Control and Information Sciences

Edited by M. Thoma and A. Wyner

Vol. 98: A. Aloneftis
Stochastic Adaptive Control
Results and Simulation
XII, 120 pages, 1987.

Vol. 99: S. P. Bhattacharyya
Robust Stabilization Against
Structured Perturbations
IX, 172 pages, 1987.